JAPANESE FLOWER CULTURE – AN INTRODUCTION

This book provides a comprehensive introduction to *ikebana* and other forms of Japanese flower culture. Unlike other books on the subject, which focus on practice, the book provides both an academic discussion of the subject and an introduction to practice. It examines *ikebana* and flower culture from anthropological and sociological perspectives, analyses Japanese aesthetics, customs and rituals related to flower arrangements, and outlines *ikebana* history and the Grand Master Iemoto system. It considers how the traditional arts are taught in Japan, and links traditional arts to current issues in today's society, such as gender and class. This book also covers how to prepare *ikebana* utensils, preserve flowers and branches, and how to appreciate arrangements, placing an emphasis on acknowledging our five senses throughout each stage of the process. The book will be of interest to a wide range of people interested in Japanese flower culture – university professors and students, tourists and people interested in traditional Japanese arts.

Kaeko Chiba is Associate Professor of International Liberal Arts at Akita International University, Japan.

JAPANESE FLOWER CULTURE – AN INTRODUCTION

Kaeko Chiba

LONDON AND NEW YORK

Designed cover image: Kaeko Chiba

First published 2023
by Routledge
4 Park Square, Milton Park, Abingdon, Oxon OX14 4RN

and by Routledge
605 Third Avenue, New York, NY 10158

Routledge is an imprint of the Taylor & Francis Group, an informa business

British Library Cataloguing-in-Publication Data
A catalogue record for this book is available from the British Library

Library of Congress Cataloging-in-Publication Data
Names: Chiba, Kaeko, author.
Title: Japanese flower culture : an introduction / Kaeko Chiba.
Description: First edition. | Abingdon, Oxon ; New York : Routledge, [2023] | Includes bibliographical references and index. |
Identifiers: LCCN 2022031770 (print) | LCCN 2022031771 (ebook) |
ISBN 9781032164687 (hbk) | ISBN 9781032164694 (pbk) |
ISBN 9781003248682 (ebk)
Subjects: LCSH: Flowers–Japan–Identification. |
Plants, Ornamental–Japan–Identification.
Classification: LCC SB406.93.J3 C45 2023 (print) |
LCC SB406.93.J3 (ebook) | DDC 635.90952–dc23/eng/20220822
LC record available at https://lccn.loc.gov/2022031770
LC ebook record available at https://lccn.loc.gov/2022031771

ISBN: 978-1-032-16468-7 (hbk)
ISBN: 978-1-032-16469-4 (pbk)
ISBN: 978-1-003-24868-2 (ebk)

DOI: 10.4324/9781003248682

Typeset in Bembo
by Newgen Publishing UK

CONTENTS

ILLUSTRATIONS

Figures

Tables

Charts

Graph

Map

ACKNOWLEDGEMENTS

This book is based on our family lives as *ikebna* teachers and my traditional arts courses at Akita International University. I am grateful to Professor Akito Okada from Tokyo University of Foreign Studies and Emeritus Professor Joy Hendry from Oxford Brookes University. I owe my sincere and warmest thanks to them for guiding me with clear and inspiring thoughts and ideas. I owe a special debt of thanks to Emeritus Professor Joy Hendry for believing in me and giving me an opportunity to publish this book. I am also grateful to my assistants Anju Kinoshita, and Yuxi Guo for helping organize this book. I also owe special thanks to Kate Entwistle, Jason Collinge, and my mother, Mihoko Chiba for their invaluable help in proofreading, as well as for critical comments on my manuscript. I can never thank my interviewees enough for their kindness, for their time spent with me, and for allowing me into their lives. Finally, I owe my greatest debts to my family for supporting me and to nature which provides peace and serenity in our hearts.

INTRODUCTION

Flowers and branches are beautiful; inner peace and contentment are abundant when we appreciate the truest imperfections of nature and being in their presence. The visual beauty of flowers and branches entices us into an imperfect natural world; a world of peace and humbleness, a state of gratitude and a space for healing. Because Japanese flower arranging, *ikebana*, captures our hearts, the simple yet skillful techniques are handed down through generations over many centuries. This book explores multiple elements and meanings behind *ikebana* and Japanese flower culture from an anthropological point of view. This chapter first introduces the nature of this book. My maternal grandmother, Yamada-*sensei* and my mother, Chiba-*sesnei* were *ikebana* teachers, so *ikebana* and flowers have always been a part of my life. This first part of the chapter shares some vignettes and a brief history of my grandmother's life in order to raise some themes, which will become common in the book, and then it proposes some research questions concerning *ikebana* and Japanese flower culture.

Autobiography

One day, when I was about to help clean my children's room, my mother, Chiba-*sensei* called me to check if I needed any flower containers. I was busy looking after the children and was not so enthusiastic about it at first, but I went anyway. Then, I found a lot of flower containers and other materials that I didn't know how to use. Most flower containers were made from pottery, quite big and heavy, some of them were quite small and had unique shapes. Most of them had simple colors, dark blue, brown and black, but some of them were decorated with bright colors, like yellow and red. I knew that most of them used to belong to my grandmother, Yamada-*sensei*. I was wondering if she liked the simple colors, and how she taught *ikebana*. Since Chiba-*sensei* was an only child, we were very

DOI: 10.4324/9781003248682-1

close to Yamada-*sensei*. We lived together until I went to university, but I don't remember how she taught her *ikebana* classes. When I was a child, she was only teaching *chadō* at her house and she rarely talked about her life with *ikebana*; perhaps she was very busy with her *chadō*. Yamada-*sensei* passed away in 2017, so I asked for her story from Chiba-*sensei*.

Yamada-*sensei*

The youngest among her siblings, Yamada-*sensei* was born in Akita city. Her father was head of the local carpentry company. She often told me that she had a very hard time when she was a child. Her parents passed away when she was small, and she often had to move around to stay with various relatives. 'I was such a pitiful girl; I had to move around all the time, really...' were Yamada-*sensei*'s usual comments. Later on, she moved to Tokyo to live with her elder sister, went to nursing school, and got her nursing licence. She met her fiancé when she was working as a nurse in Tokyo; he was a medical doctor. Chiba-*sensei* told me that Yamada-*sensei* was asked by his mother to learn *ikebana* and *chadō*, so she could become his wife. That was how Yamada-*sensei* started to learn *ikebana*, Ichiyo School of *ikebana* (Figure 0.1 and 0.2). However, her fiancé died in the war and she never was able to marry him.

FIGURE 0.1 *Ikebana* practice in the 1940s in Tokyo

Source: Photograph by a person unknown.

FIGURE 0.2 *Ikebana* practice in the 1940s in Tokyo

Source: Photograph by a person unknown.

I found an old suitcase full of my grandparents' pictures. I vividly remember finding a picture of an unknown young man in his military uniform. He did not look at all like my grandfather nor his relatives, so I asked my grandfather who he was. He mumbled, 'Oh, he is your grandmother's ex-fiance…' He looked very young and naïve, but kind. Later in her life, my grandmother, Yamada-*sensei* asked my mother, Chiba-*sensei* to take her to Mutsu Bay in Aomori. Yamada-*sensei* told Chiba-*sensei* that she came to see him, as he was apparently serving on a military ship as a medical doctor. It was the last time for her to see him. Later, Yamada-*sensei* also asked Chiba-*sensei* to take her to Osorezan in Aomori Prefecture to communicate with her ex-fiancé. Osorezan is well-known as the place where it is believed that spiritual mediators can reach beyond the veil to communicate with people who have passed away. Yamada-*sensei* was desperate to talk to him once more.

After the Second World War, Yamada-*sensei* came back to Akita city, and she worked for a local elementary school as a school nurse. There, she met my grandfather and got married (Figure 0.3). She stopped working as a nurse and started

FIGURE 0.3 Yamada-*sensei* in 1940s

Source: Photograph by a person unknown.

teaching *ikebana* at her house. My grandfather was working for a local newspaper company, and they moved to the northern part of Akita called Noshiro city, which is a smaller town than Akita. Chiba-*sensei* remembered that Yamada-*sensei* often taught *ikebana* to young single women. She recalled that there were about ten women in her class. Chiba-*sensei* commented,

> At that time, it was very common somewhere in the Noshiro or Hanawa area for ex-land owner's girls to do bridal training (*hanayome shugyō*), and your grandmother was quite busy teaching these girls. Whenever I got back from school, they were in the *tatami* room practicing *ikebana*.

When I heard my grandmother, Yamada-*sensei*'s story, I felt that without her commitment to her ex-fiancé, I might have never opened my door to *ikebana*.

My mother, Chiba-*sensei* told me that she grew up with *ikebana*; she remembered that there were always young women practicing *ikebana* at her home and sometimes she practiced next to them (Figure 0.4). When Chiba-*sensei* was asked to teach an *ikebana* course at a university, she never hesitated. While I was looking at old pictures, my grandfather told me a mysterious story. My mother's grandfather's grandmother did not have any male siblings and had to arrange a marriage to have a husband enter the family. Her first husband was so into *ikebana* and other artwork that he did not work hard in his full-time job with his

FIGURE 0.4 Yamada-*sensei* in 1960s

Source: Photograph by Minoru Yamada.

company. He was divorced by her family. My grandfather showed me the script written and drawn by him. He drew the *ikebana* arrangement and included a detailed description. It was in black ink, but his drawing touch was so delicate that I could imagine what he had arranged.

The start of my *ikebana* journey

My mother, Chiba-*sensei* started to learn *ikebana* from Sugidate-*sensei*. With Chiba-*sensei*, I practiced a different style of *ikebana* arrangement, *suimenka*. This was the arrangement whereby hydrangeas were cut rather short and placed close to the water level. Chiba-*sensei* once selected a dark-blue colored container with flowers which were light purple, pink, and white in color. It created a moment to feel the coolness from the water, the deep color from the container and the harmony of flowers matched so smoothly. It looked as if there might be someone standing right underneath that hydrangea, reminiscent of a little child waiting underneath the tree on a rainy day. Hydrangeas did not have a strong fragrance,

FIGURE 0.5 Hydrangea

Source: Photograph by the author.

but I felt the scent of rain soaking into the soil and its greenness. I felt peace wash over my mind just looking at the water in the dark-blue container, hydrangeas, and dewdrops on the leaves (Figure 0.5).

I also started to learn *ikebana* from Sugidate-*sensei* twice a month. I felt I had to practice *ikebana* if I was to start teaching it. Even though it was part of my work, I felt it was a very luxurious time (*zeitaku na jikan*). I often had a quiet moment with my flowers and branches so as not to rush examining where to place them. Whenever I thought about this matter, there was often a moment of struggle, feeling oh no this is difficult, but there was also a moment to feel the sense of calmness by touching nature. It was a private lesson and for about 90 minutes, flowers and branches entertained me, and they had my utter focus. The flowers were selected by Sugidate-*sensei*. She prepared a variety of flowers, but often I did not even know their names; the names were very difficult to remember. Several months later, I was advised to practice *rikka* style. There were around nine branches this time and I felt the order of the arrangement was confusing, but what was more surprising was that *ikebana* allowed me to use wires to bend

FIGURE 0.6 Rikka

Source: Photograph by the author.

branches up to 90 degrees (Figure 0.6)! I was shocked to learn that *ikebana* used any artificial method at all and also did in fact encourage a certain degree of unnaturalness. Sugidate-*sensei* commented, '*ikebana* indeed, uses a lot of wires, staples, tapes, you think *ikebana* is all natural but it is not.'

There were some *ikebana* exhibitions, practitioners including my *sensei*, arranging flowers, and there were a variety of flower arrangements, from classical setups to the free styles. All of the practitioners looked serious and arranged their work while they were preparing. But once it started, practitioners who had arranged works greeted guests with smiles and seemed to share pleasant moments (Figure 0.7). The majority of the guests and audience were senior women over seventy. The audience generally came with friends and they tended to chat about the flower arrangements, sometimes commenting, 'Wow, this is an unusual flower. I wonder what it is! Have you seen this?' There were flower arrangements which also used other materials; plastic bottles, sneakers, or even the Winnie the Pooh character. Some other times, there were arrangements without any natural materials (Figure 0.8). One of the *ikebana* teachers commented, 'Please

FIGURE 0.7 *Ikebana* exhibition

Source: Photograph by the author.

FIGURE 0.8 *Ikebana* exhibition without flowers

Source: Photograph by the author.

look at the art work as one of the works with others. These artworks without any flowers are also considered in our *ikebana* school.' I had taken it for granted that flowers should always be with the arrangement, and felt *ikebana* was more creative than *chadō*, which I also practiced. Since my grandmother and mother were actively engaged with *chadō* as teachers, I grew up with *chadō*: the smell of *matcha*, the sound of the kettle, the taste of an unsweetened sweet, and the simple *tatami* floor. More precious than anything else though, I grew up with *chadō* practitioners around me. As I started to practice *ikebana*, I felt that *chadō* maintained a stricter atmosphere.

I had an opportunity to visit one of the *ikebana* grand master (*iemoto's*) houses. I thought I was going to meet only her, but she had two other women sitting next to her. It seemed that there was no privacy, but later I realized that they were *iemoto* assistants. When she explained that it was her practice room (*keiko-ba*), she also explained that it was her house. It appears that it is common for many teachers to offer their practice in their own homes. The *iemoto* shared her stories of how she had learned her *ikebana*. She had rarely learned from her father. Instead, she had learned from her sister's pupils. She further told me that her biggest fear was that *ikebana* culture was dying out. She mentioned, 'At the peak time in Akita city, we had 20 schools and around 800 practitioners. Now it has decreased to 16 schools and only around 300 practitioners, and most of them are older generations.'

Once when I was watching TV to relax, there was an advertisement showing a scene of several women in their 30s–40s taking a western flower arranging lesson in a private home. It expressed the elegance of *ikebana*, but it was not Japanese *ikebana*. The audiences of Japanese *ikebana* and western flower arranging were quite different. My worries also led me to question: is *ikebana* dying? Similar worries also arise when I go walking around my neighborhood. It is indeed beautiful to walk around the residential area in Akita during springtime, seasonal flowers, plum trees, cherry blossoms, and daffodils can be seen in many gardens. However, there are hardly any houses which have Japanese-style gardens with stepping stones, a stone lantern, and pine trees with moss. Most of the houses around Akita are made with concrete in the style of western houses. Most of them also have western-style gardens with neat flower beds and roses which stand clear of weeds and nature's imperfections. Are we also going to have fewer and fewer Japanese gardens in Japan?

So far, I have recounted some stories detailing the influence of *ikebana* and flower culture within the history of my family's relations and leisure time. The following questions have arisen during this brief and personal reflection.

Question 1

What was it like to have *ikebana* lessons during the Edo period? Did my ancestor learn by himself? With what kind of flowers did he practice?

Question 2

Can *ikebana* artwork without any flowers be considered real *ikebana?* Do all the *ikebana* schools accept this kind of artwork? How do the styles differ from other schools?

Question 3

What is it like to be *iemoto*? In what ways are other traditional arts schools accepting female *iemotos*?

Question 4

Are Japanese *ikebana* and Japanese garden culture likely to gradually die out in the future?

Question 5

Why did many young women want to practice *ikebana*? Why do many flowers tend to be connected to femininity? Did women really want to practice it or were they advised by their family members? Why do senior women still want to practice *ikebana*?

Meaning of this book

This book aims to introduce *ikebana* and Japanese flower culture from a social anthropological perspective. Each chapter shares some topical discourses, analysis, and interview data, and also provides research questions and references at the end. The traditional arts way of learning is related to Zen Buddhism and does not encourage asking questions. However, this book is targeted at university students, teachers, and researchers, and as such, I have suggested some critical questions at the end of each chapter. I hope these give ways to inspire more in-depth explorations and debates, and help keep alive the discourse around *ikebana* today.

This book is based on fieldwork I carried out in Akita City over five years, 2015–2020. I attended regular *ikebana* lessons like the other practitioners and gathered information. During the fieldwork, I also conducted around thirty interviews with *ikebana* practitioners, their family members, teachers, *iemoto*, and artisans. This book pays great emphasis to these direct voices among other elements. The majority of my informants' voices are not publicly heard due to their social positions as women based in non-metropolitan areas. Still, the majority of academic debates regarding gender, social class, education systems, and human rights in Japan tend to be founded on data collections based around metropolitan areas due to the lack of academic institutions in the non-metropolitan areas of

Japan (Mock 2016). This book shines a light on these unheard voices and unrecorded documents, and hopes to contribute to the further discussion and analysis of gender studies, social class issues, and traditional arts studies.

The descriptions in this book are based on my numerous experiences from *ikebana* practice. Proust (1941) describes in his novel *Remembrance of Things Past* that explaining a single event is not an easy task for us. When we describe one specific church, we have to have seen other kinds of churches numerous times. Observing the various shapes of stained glass, roofs, alcoves, doors, and windows in other churches makes us naturally see the uniqueness of the church and able to give a detailed description including the atmosphere of the church. Proust points out the critical evaluation of a person's memory and explains that memory is constructed from a person's countless similar experiences. Indeed, my memory of *ikebana* and flower culture in Japan is inevitably shaped by repetition, and descriptions in this book are based on these recollections.

My analysis of *ikebana* and flower culture is based on my standpoint as a Japanese woman raised primarily in a non-metropolitan area, Akita; a granddaughter and a child of an *ikebana* teacher, a mother of two who wishes one of them carry on what I do, an anthropologist educated in the UK, and ultimately a teacher who makes every effort to share the meaning of Japanese traditional arts. The approach to the *ikebana* analysis is inevitably different from other researchers who have their own unique personal backgrounds. I examined it from my standpoint, as a native anthropologist, and I hope this book also shares other perspectives toward understanding *ikebana* and flower culture.

My standpoint might be criticized as being too close or subjective. I argue with Cancian (1992), Harding (1987), and Smith (1987) that social science itself does not hold an objective point of view. Although dominant intellectuals have claimed that their research standpoint is 'neutral,' 'universal,' and 'objective,' their knowledge or motivation is always influenced by their own genders, social statuses, and racial backgrounds. I argue that expressing the clearest background and standpoint of a researcher should make their research more understandable and relatable. With practitioners' direct voices, the following is the flow of this book.

Structure of the book

The Japanese are very conscious of seasonal changes. Chapter 1, 'Japanese decoration,' rituals and symbols looks at how flowers and branches are used for Japanese events and rituals throughout the year and throughout a person's life. This chapter explores which flowers are appropriate and which are taboo for specific occasions. It will further examine the meaning behind these seasonal flower arrangements.

Chapter 2 analyzes 'Japanese aesthetics': the concepts of *wabi sabi* and *miyabi*. By sharing examples of *wabi sabi* in Japanese traditional arts and culture, this second chapter will also contrast them with aesthetics in other countries. Readers will further explore how other artists have been influenced by the concept of *wabi sabi* outside of Japan.

Japanese aesthetics and *ikebana* have changed throughout time. Hence, Chapter 3, 'History,' introduces how these aesthetics, the flower culture, and *ikebana* have evolved. It examines how flower arranging in Japan began with a religious influence and gradually changed with the architectural styles of the time. It will also proceed to explore how *ikebana* has changed in relation to gender roles in Japan.

Most of the traditional arts are maintained through the *iemoto* system. Chapter 4, '*Iemoto* system,' focuses on the characteristics of this system including the licensing regulations. It also touches on the current discourse regarding gender within the *iemoto* system. Readers will further analyze the different schools in the *ikebana* field and other traditional arts, too including Japanese dance (*nihonbuyō*) and *noh theatre*.

Sen no Rikyū is recognized as the founder of *chadō*, and he arranged the flowers as if they were blooming in a field. The style is quite simple, yet it conveys significant messages. In Chapter 5, 'Tea flower,' readers will take a closer look at how the tea flower arrangement is different from an *ikebana* arrangement in addition to the importance of appreciating the five senses. This chapter also concentrates on what kinds of flower containers are appropriate for specific occasions.

There are different styles of Japanese garden (*nihon teien*) in Japan. Chapter 6, 'Japanese garden,' describes how flowers have been utilized to create the recognizable Japanese garden. It also examines the in-depth meaning of each type of garden and how numerous types of techniques are established to maintain beauty in Japanese gardens. This chapter will share an interview from a highly skilled gardener (*niwashi*) to reveal their training style and work ethic.

Flowers are not only used for *ikebana* arrangement, but also in literary works to express a person's inner feelings, spirits, beliefs, and messages to others. In Chapter 7, 'Literature,' readers will explore how the flower has been used in various kinds of literature from ancient up to modern times in *Kokin Wakashu*, *The Tale of Genji*, *The Pillow Book*, *Takekurabe*, as well as through the language of flowers (*hanakotoba*). In this chapter, readers will also examine how women's experiences are evoked in flowers in both *The Tale of Genji* as well as a related modern work, *Onnamen (Masks)* by Enchi Fumiko.

Ikebana was once only accessible by men, and only after the Meiji period was it officially made available to women. Chapter 8, 'Gender and class,' investigates how *Ikebana* influences, and is influenced by, the gender and class customs which are part of Japanese society. A sociological approach helps us to identify the issues arising from these traditional arts, and engage personally with individuals' contributions to the anthropological understanding of the social context. Readers will explore the reasons why senior Japanese women still want to practice *ikebana* even after their bridal training (*hanayome shugyō*).

Traditional Japanese arts, and the *ikebana* teaching style, are heavily influenced by Zen Buddhism training: simple observation, imitation, and repetition. Chapter 9, 'Traditional art education' begins by describing how *ikebana* and other Japanese traditional arts have been taught mainly in Japanese compulsory education. It then proceeds to analyze how this traditional art education differs according to varying teaching styles and regions in Japan. In order to explore effective teaching styles, the chapter shares comments from teachers and those who engage with traditional arts education. It also highlights some challenges to expanding *ikebana* in the current education field.

There are numerous styles of *ikebana* arrangements depending on the occasions and seasons. In *ikebana*, practitioners not only emphasize flower materials, but also the flower containers (*kaki*). Chapter 10, 'Utensils and *ikebana* arrangements,' will have a closer look at *ikebana* utensils and basic techniques. Readers will learn how to cut and preserve flowers and branches. There are also many types of *ikebana* arrangements, styles, and regulations, which are different from school to school. For readers to enjoy the actual practice, this chapter guides the reader through the basic arrangements including *saika*, *moribana*, *nageire*, *shinseika*, and freestyle, along with some ideas to conduct online *ikebana* practice and field trips.

The last chapter draws together the themes of the previous chapters and concludes this book by introducing an additional approach for exploring *ikebana* and flower culture, including collaborations with digital and pop cultures. It examines *ikebana* and other traditional arts' futures by rethinking their history, the remaining essence of nationalistic sentiments, the current status, and gender issues. Now, please turn the page and explore the beauty of flowers and the hidden meaning behind them!

References and further reading

Cancian, F. (1992) 'Feminist science: Methodologies that challenge inequality,' *Gender & Society*, 6, 623–642.

Harding, S. (1987) *Feminism and Methodology*, Bloomington and Indianapolis, IN: Indiana University Press; Milton Keynes: Open University Press.

Mock, J. (2016) 'Smart city - stupid countryside: Social and political implications of the urban/rural split in Japanese education.' In Mock, J. Hiroaki K and Naeko N (eds.), *The Impact of Internationalization on Japanese Higher Education: Is Japanese Education Really Changing?* Rotterdam: Brill, pp.191–206.

Proust, M. (1941) *Remembrance of Things Past, 5*, trans. C. Scott-Moncrieff, New York: Random House.

Smith, D.E. (1987) *The Everyday World as Problematic: A Feminist Sociology*, Boston: Northeastern University Press.

1

DECORATIONS, RITUALS, AND SYMBOLS

The Japanese are very conscious of seasonal changes. This chapter looks at how flowers, branches and *ikebana* related arrangements are used for Japanese events and rituals throughout the year and a person's life. Readers will explore which flowers are appropriate and which are taboo for specific occasions, they will also learn the meaning and wishes behind these flower arrangements. The chapter further discusses how these arrangements have developed to accommodate the needs of modern life which emphasizes efficiency (Cross 2009, Hartmut 2013). While some readers might feel that this content is common knowledge, it is also worth noting that barriers such as one's social class, region, and generation will inhibit the diffusion of common knowledge in Japanese society. Thus, it is worth reviewing these introductory topics before delving into in-depth discussions of *ikebana* culture. First, let us explore the beauty of annual flower arrangement events that are regular and habitual in daily Japanese life.

Yearly events

January

1/1 New Year, kadomatsu *– bamboo*

For celebrating the New Year, two *kadomatsu* are arranged at the front of the entrance hall (*genkan*). It is believed that a *kadomatsu* is the sign which guides the New Year god (*toshigami*), every year from the mountains to the family home to protect the household. As Figure 1.1 shows, three freshly cut bamboos are wrapped with straw (*wara*). The bamboo changes color once it is cut and turns a light brown color as it ages. The green bamboo is used in *kadomatsu* to express the fresh new start to the year. The *kadomatsu* arrangement is related to Shintoism

DOI: 10.4324/9781003248682-2

FIGURE 1.1 *Kadomatu*

Source: Photograph by the author.

(Kawase 2000). Tall evergreen trees are similarly used as antennas to attract gods; these trees are called *yorishiro*.

Shō chiku bai – *pine trees, bamboo, Japanese apricot*

These three decorations are used for celebrations including New Years, graduations, and weddings (Figure 1.2). They are also used as decorations for products such as *sake*, sweets, and noodles. Pine trees as evergreen trees express longevity. Bamboo represents strong vitality from the fact that it is hard to bend and grows fast. The Japanese apricot is understood as a symbol of nobility and longevity from its character of providing beautiful flowers even as it ages (Tsutsumi 1997).

February

2/3 Setsubun – *hollytree*

Setsubun literally means the division of seasons. It is recognized as the start of spring, *rishun*. To bring more fortune to each household and elsewhere, beans, believed to have power to get rid of bad spirits, are thrown to an imaginable ogre figure as one utters, 'Ogre is outside, fortune is inside (*Oni ha soto, Fuku ha uchi*).' Some households arrange the head of a fish pierced on top of a holly tree (*hiiragi*) (Figure 1.3). The pointy leaves on the holly tree plant are believed to stab bad ogres' eyes, and the strong smell of fish heads is to get rid of bad spirits. They are placed at the front of the entrance hall (Saito 2018, Shintani 2019).

FIGURE 1.2 Pinetree, bamboo, Japanese apricot decorations for a New Year money envelope

Source: Photograph by the author.

2/14 Valentine's Day – chocolate, not roses

This custom became popular in Japan from the 1960s – considerably influenced by chocolate companies. While offering beautiful roses might be a well known ritual in the West, offering other gifts, especially chocolates seems to be more common in Japan (Shintani 2019). This custom was clearly divided by gender: On Valentine's Day women are supposed to give gifts to men including their boyfriends, husbands, their colleagues, and bosses. One month later, on White day, men are supposed to give some gifts back to them. However, it has become more common to give gifts regardless of gender, some people even buy chocolate for themselves.

FIGURE 1.3 Holly trees for *setsubun*

Source: Photograph by the author.

March

3/3 Girl's Day festival (hinamatsuri) – peach blossom flower (momonohana)

Wishing for girls' happiness, *hina* dolls are specially decorated. Peach blossom flowers are arranged next to *hina* dolls, as it is believed that they have powers to get rid of bad spirits (Figure 1.4). There is a famous song for *hinamatsuri* in which the lyrics state, 'Let's arrange the peach flowers (*Ohana wo agemasho momono hana*).' Due to this popular song, it seems to be well-known that the peach blossom flower is appropriate and often sold during *hinamatsuri* in Japan. For *hinamatsuri* occasions, specific sweets, including diamond-shaped mochi (*hishimochi*) and rich crackers (*hinaarare*), are prepared. Both sweets are made using colors which represent springtime, including light green to symbolize the new fields, pink to represent the flowers, and white to indicate the remaining snow (Saito 2018). Saito (2018) also argues that pink, white, and light green represent talismans, purity, and good health respectively. It is also said that the dolls should be removed by the house owners as soon as possible after the day, otherwise the

FIGURE 1.4 Peach flower arrangement

Source: Photograph by the author.

girl will not get married! This might be an interesting question in relation to gender issues. How long will this tradition live in an era that is growing out of commenting on women for not getting married before it fades into history?

April

Cherry blossoms viewing (hanami)

Around this time of year, the Japanese enjoy cherry blossom viewing. They gather underneath the cherry blossom trees and enjoy a picnic of food and drink (Figure 1.5). Some famous viewing places open shops, and police navigate the drunken guests. Television broadcast companies offer daily forecasts detailing the best locations to enjoy the blooming blossoms. *Hanami*, literally translated as 'flower viewing,' is only used for cherry blossom viewing. The existence of this festival shows how attentive the Japanese are to the change in seasons and yearly occurrence of the cherry blossoms. It is said that this *hanami* custom has derived from the belief that a God exists in this type of tree, and communal praying took place underneath the trees with *sake* and fresh food (Kawase 2000). *Mochi* with three different colors (*hanami dango*), are offered around the same time as *hanami* season. Much like the colors of the sweets used for the girls' day festival, the same pink, white, and light green are used for *hanami dango*.

FIGURE 1.5 Cherry blossom viewing

Source: Photograph by the author.

May

5/5 Children's Day (kodomo no hi) – iris

Large fish-shaped flags, known as *koinobori*, are used outside the house to celebrate *kodomo no hi*. Japanese armor, swords, and archery arrows decorate the home. This day was originally for celebrating boys, hence the decorations are stereotypically and traditionally masculine, bold colors and images of bravery, courageousness, and strength. Irises (*shōbu*) are arranged next to dolls and decorations. This flower is considered the masculine flower in Japan because the shape of the leaves is reminiscent of swords. During the evening of this day, some families appreciate irises not only visually, but by using the scent of the leaves to create an aromatic bath. It is believed that rubbing iris leaves on the skin can get rid of bad spirits. They have a minty scent which makes people feel refreshed. For this day, *chimaki*, *mochi* wrapped with bamboo leaves, are served wishing to get rid of bad spirits. *Mochi* wrapped with oak tree leaves, *Kashiwa mochi* is also served wishing for longevity; this custom came from the nature of the oak tree leaves because they only begin to fall once the young leaves start to appear (Tsutsumi 1997). While *chimaki* is generally arranged in western Japan, *Kashiwa mochi* is well known for this day in eastern Japan.

FIGURE 1.6 Iris for Children's Day

Source: Photograph by the author.

FIGURE 1.7 Mother's Day advertisement

Source: Photograph by the author.

Mother's Day (haha no hi) – *carnation*

Just like other countries, Japan also celebrates Mother's Day. Carnations tend to be the common flowers to send (Figure 1.7). The colors of the flowers are carefully chosen; white-colored carnations are sent only for mothers who have passed away since this color connotes death as well as purity. Mother's Day and the custom of giving carnations were influenced by the USA during the Taishō period (AD 1912–1926) (Saito 2018).

June

Rainy season (tsuyu) – *hydrangea*

It is not a special day or season to celebrate, but this rainy season always arrives every year. Hydrangeas are commonly seen in this season and used for greetings in letters, postcards, stamps, poems including haikus, decorations, and *ikebana* arrangements (Tsutsumi 1997) (Figure 1.8).

FIGURE 1.8 Hydrangea

Source: Photograph by the author.

July

7/7 Star festival (tanabata) – bamboo

Usually, children write their wishes on a small piece of card, which is hung and displayed on bamboo and is believed to have the power to get rid of bad spirits (Figure 1.9). As with New Year's Day, bamboo is commonly known as the decoration for this specific day. The lyrics in a star festival say, 'The bamboo leaves are blowing through the breeze. Stars are shining in the sky.' Bamboo grows naturally in Japan, and it is not difficult to get bamboo from around the neighborhood in a non-metropolitan area. Bamboo is not only used as dishes in which cold noodles are served during the Tanabata seasons, but also as utensils, materials for houses, and *ikebana* materials (Saito 2018).

August

8/13–16 Obon festival – chrysanthemum (kiku)

Ancestors are believed to return to the family homes during this period; the Japanese visit their family graveyards and clean and decorate the family alcove at home to welcome their ancestors' return. For this occasion, chrysanthemums are

FIGURE 1.9 Children decorating for *tanabata*

Source: Photograph by the author.

used for decoration at the graveyard and for family alcoves (Figure 1.10). Since it is the most common flower used for *obon* and funerals, the chrysanthemum is a symbol representing death. However, it actually has a positive meaning as it signifies honour, respect, and purity. Consequently, it is the flower especially dedicated to the Imperial Japanese family (Tsutsumi 1997).

FIGURE 1.10 *Obon* arrangement for family alcove

Source: Photograph by the author.

September

9/9 Chōyō no sekku – *chrysanthemum*

Originally derived from China, odd numbers have been considered as fortunate numbers. Therefore, 7 January, Girl's Day, Children's Day, *tanabata*, and this day are considered five ceremonial days (*gosekku*). The chrysanthemum is generally arranged for this day, as it is believed that having *sake* with chrysanthemums (*kikuzaku*) while admiring chrysanthemum flowers brings longevity (Saito 2018).

FIGURE 1.11 Pampas flower arrangement

Source: Photograph by the author.

Moon festival (otsukimi) – pampas grass (susuki)

During September, a small collection of offerings is placed in the evening under the light of the moon. To wish for a good harvest and show appreciation for the harvest, some food and sweet rice dumplings, *dango*, are served, along with a small arrangement of pampas grass (Figure 1.11). Pampas grass resembles ears of rice, *inaho*. Rice itself is perceived as a precious food, which is highly respected by the Japanese. Children are taught to not leave a single grain of rice in their bowls in order not to waste any, and to show gratitude to the farmers who made it. In summary, the high respect for rice within Japanese culture is symbolized by the offering of Pampas grass to the gods during *otsukimi* (Kawase 2000, Tsutsumi 1997).

Seven Autumn flowers – Aki No Nanakusa

Pampas grass (*susuki*), Bush clover (*hagi*), Arrowroot (*kuzu*), Pink flower (*nadeshiko*), Dahurian Patrinia (*ominaeshi*), Boneset (*fujibakama*), and Balloon flower (*kikyō*) are the seven most well known autumn flowers and are often seen in generic flower, *chadō* flower, and *ikebana* arrangements. Students are encouraged to learn these seven flowers from a young age from a Japanese language textbook. The pink flower is also known as a representation of Japanese beauty, so the Japanese women's soccer team is named Nadeshiko Japan. Arrowroot is used as an ingredient in Japanese sweets, such as *kuzu mochi* and *kuzukiri* (Shintani 2019).

October, November

Kōyō *(colored leaves)* – momiji

Much like how tourism surges during the cherry blossom season, many Japanese take part in national tourism to visit and appreciate the beautiful colors of autumn, some travelling especially far. Tourist companies arrange multiple tours to visit beautiful and scenic hotspots for day trips and small getaways. Weather forecasts also provide the best locations to view the strikingly colorful leaves. The change of color in the leaves as autumn approaches and the blooming of the cherry blossoms in the spring are clear examples of how locations across Japan experience seasonal changes at different times of the year; in Hokkaido, the leaves begin to change color around the beginning of October, while in Fukuoka in Kyushu, they only start to turn around the end of November (Shintani 2019).

December

Christmas – holly

While Christmas is not originally a Japanese custom, the majority of Japanese, especially those who have small children, tend to celebrate this event. As in other countries, poinsettia and holly leaves are used as decorations (Figure 1.12). Artificial trees are the most common choice for the indoor Christmas tree in Japan, rather than the pine tree, which may be commonly used in other countries (Saito 2018).

FIGURE 1.12 Poinsettia

Source: Photograph by the author.

So far, the flowers used during the yearly celebrations have been outlined. The seasonal flower arrangements are also often reproduced in *ikebana* arrangements. Let us now explore what kinds of flowers are used for life events.

Graduation ceremony

Most students in Japan graduate in March. For this occasion, some students might receive a small flower bouquet as a congratulatory gift. Flowers in Japan are relatively expensive, so it may be common to see a senior student (*senpai*) receiving a few tulips, Baby's Breath (*kasumiso*) arrangements from the junior students (*kohai*) belonging to the same sports clubs.

1/15 Coming of age ceremony (seijinshiki) – flower arrangement

Japanese municipalities arrange *seijinshiki* gatherings to celebrate becoming an adult. On this occasion, girls generally wear a long-sleeved *kimono* (*furisode*). On the stage, a beautiful flower arrangement is placed to express the celebration wishes; this used to be an *ikebana*-style arrangement, but has gradually changed to western flower arrangement styles due to its modern image.

Weddings

It appears that *ikebana* or other flower arrangement styles were not commonly used for weddings in Japan. When the western-style wedding became popular from the 1960s onwards, so too did the style of western flower arranging (Saito 2018). Nowadays, different kinds of flowers tend to be used for weddings, except the typical large chrysanthemums, which still remain flowers dedicated to funerals.

Hospital flowers

If a person is ill, it is appropriate to avoid potted flowers (Figure 1.13). There is a saying in Japanese that sickness stays in the root (*byoki ga netsuke*). Flowers with a strong fragrance, including rose, lavender, and lilies are not recommended since some patients are sensitive to smells. Additionally, some flowers have death connotations and should not be sent. Camellias are often avoided because when they fade, the entire flower head tends to fall off. This resonates with some Japanese who hesitate buying this flower due to its resemblance to decapitation (*kubikiri*) (Shintani 2019).

FIGURE 1.13 Flowers for patients

Source: Photograph by the author.

FIGURE 1.14 Chrysanthemum design for coffin

Source: Photograph by the author.

Funeral

Flowers are often used for offering condolences. For the arrangement style, it was more common to see the wreath (*hanawa*) style rather than *ikebana* at a funeral. This *hanawa* style could also be seen for celebrating the opening of new businesses, for which the decorative color changes from black and white for a funeral to red and white (Saito 2018). Due to the cost of flowers and convenience, some businesses used plastic *hanawa* in front of their shops. Some funerals offer *hanawa* made out of towels so that they can give them as gifts. To express respect to the person who has passed away, the most honorable flower, chrysanthemum, is often used for designs to cover the coffin, as in Figure 1.14. For the funeral ceremony, not only chrysanthemums, but lotus designs, too, are arranged as the sacred flowers of Buddhism symbolizing enlightenment (Figure 1.15).

National, prefecture, city, and family symbols

Flowers are not only symbolic of a specific year or life event, but also of the nation, prefecture, city, and family household in Japan (Table 1.2). While there aren't any flowers which are officially registered as the national Japanese flower, Japanese dictionaries including *Kojien* (Shinmura 1991) state that the

FIGURE 1.15 Lotus flower for funeral

Source: Photograph by the author.

chrysanthemum (*kiku*) and cherry blossom (*sakura*) are both national flowers decided by custom and nation (Shintani 2019). *Kiku* has been used as the symbol of the Imperial Japanese family, Japanese embassies, and police stations. Japanese passports also have chrysanthemum images. Each prefecture and city has flower and tree symbols.

TABLE 1.1 Flowers for girls' name

Name in Japanese	Meaning
Sakura	Cherry Blossom
Ume	Japanese plum tree
Ayame	Iris
Himawari	Sunflower
Ran	Orchard
Fuji	Wisteria
Sumire	Violet
Yuri	Lily
Botan	Peony

TABLE 1.2 Prefecture flower

Prefecture	Flower
Aomori	Apple flower
Tokyo	Cherry blossom (*Somei yoshino*)
Fukui	Daffodil
Mie	Iris
Kyoto	Cherry blossom (*Shidare sakura*)
Yamaguchi	Mandarin flower (*Natsumikan*)
Fukuoka	Japanese Apricot flower
Kagoshima	Gentian (*Rindo*)
Okinawa	Erythrina Variegata *(Deigo)*

Flowers, leaves, and trees are often used as the symbols of a family as a family crest (*kamon*). For instance, the triple hollyhock mark is well known for Tokugawa families who ruled Japan during the Edo period (AD 1603–1868) (Lebra 1995). While well-known English surnames: Smith, Taylor, and Millar are related to occupation, Japanese surnames: Suzuki, Sato, Ito, Kato, Sasaki are related to plants and appear to be common in Japan. Flowers tend to be used for girls' names wishing for their beauty, wisdom, and happiness. Japanese apricot (*ume*), cherry blossom (*sakura*), and chrysanthemum (*kiku*) were used for a long period of time. Table 1.1 displays other examples. On the other hand, names such as Hideki, Naoki, and Daiki are associated with trees (*ki*), and have been more popular for boys since the Second World War (Sasaki 2003). Sasaki (2003) argues that names have become gender-neutral after the millennium, for instance, the name Ren (lotus), has become popular for both boys, girls, and non-binary, or gender non-conforming people. Lotuses grow in the swampland and support beautiful flowers, so Ren means purity and holiness that will endure hardship. As commented above, most flowers are related to feminine images, except the iris flower in Japan.

Dying out? Reviving? Is it really Japanese?

Yanagida (2019) argues that Japanese life is constructed with non–daily (*hare*) and daily (*ke*) concepts. Yearly events, such as New Year celebrations, *setsubun*, girl's days, children's days, and weddings are understood as *hare*. Segawa (1982) argues that funerals are also perceived as *hare*. Flowers are used in cuisine for *hare* occasions or to decorate and purify. As Cross (2009) and Harmut (2013), commenting on the issues in relation to our modern life prioritizing efficiency, the culture around flower-related decoration seems to have become simple in Japan. It had been common for women to situate their non-paid labor within their homes, allowing the knowledge and art of floral decoration to be taught to each new generation. This has helped to keep this traditional culture alive; however, nowadays women are much more likely to take up employment outside of the home, which potentially puts the cultural heritage of floral decorating at risk. Additionally, it appears that the increasingly common structure of a nuclear family in Japan has significantly affected these traditional events. Most nuclear families live in small houses, which do not have enough space for large quantities of decorations, such as dolls or large flower arrangements. Grandparents or other relatives used to help with traditional decorations, however, this help seems difficult to come by due to the new Japanese lifestyle: living as a nuclear family.

According to Watanabe (2011), less than 30 percent of Japanese claim that they believe in religion. Suzuki (2006) states that although not many Japanese claim that they believe in religion, 66.9 percent of Japanese answered that they actually follow the custom. On the other hand, a survey was conducted to explore how the younger generation, aged 18–24 years, engaged with traditional Japanese events in 2019. Around 85 percent of students have followed these traditional events, but in a simple way. For instance, they commented that their parents or mainly mothers decorated small sized *hina* dolls. An interview was conducted with a doll-making company in Akita city, Akita Ningyo Kaikan. Akita Ningyo Kaikan (2021) commented that the size of *hina* dolls has become substantially smaller than in previous generations:

> You see, nowadays, people prefer to have much smaller ones, like the size you can place at the corner on the shelf. Not many people have *tatami* floors where we can occupy the entire rooms with *hina* dolls. They are happy to pay money for dolls, so they spend money, not for the size, but for the quality. For example, they spend money for a specially ordered one instead of a regular one. In the Showa era, everything was big, big *hina* dolls, big *koinobori*, but now many things are small and simple.

Indeed, while the large *koinobori* for Children's Day may be seen in the public space, it is rarely seen outside of the individual household. Instead, small-sized *koinobori* offered at the 100 yen shop are often seen in residential areas. A similar phenomenon appears to be seen in the Japanese family alcove (*butsudan*) industries. Yame city in Kyushu is well known for *butsudan* production. A local alcove artwork collaborator from Unagi no Nedoko, Watanabe (2021), comments that *butsudan* have become much smaller and have a different design to adjust to the modern home conditions in Japan.

This simple ritual is relevant to weddings and funerals. In rural farming communities where there is less entertainment, weddings, and funerals were considered special events for local people to gather at. Until the 1970s, these rituals were sometimes held for a week in the northern part of Akita city (Kitaakita City Board of Education 2010). Most funerals have been related to Buddhist temples; some of the families in Akita still pay around 3 million yen (26.8K US$) for funerals and related rituals. However, there have been numerous cases where simple funerals from 100,000 yen budget are held (Shimada 2010). Goto (2021), who currently works for one of the largest funeral companies in Akita city comments:

> Nowadays, it is very common to do family funerals even if your family is not poor. People do not see any point in spending money and their time for funerals. Rather, they comment that they want to spend money for traveling, hobbies, … less time and effort for funerals.

On the other hand, kindergartens, day care centers, elementary schools, and the public domain provide substantial opportunities to engage with these events. Most preschools provide not only events, but also monthly books, which describe the origins and the meanings of the events. Japanese language textbooks in elementary schools provide seasonal flower-related information. If we consider these efforts to share Japanese flower culture, it appears that such events and campaigns have been accessible to everyone in Japan, but is it really so? These decoration customs are much more commonly practiced by those privileged by being middle or upper-middle class, and only relatively well-off families in Japan are able to purchase *hina* dolls and *satsuki* dolls for Children's Day (Lebra 1995). Another question arises, does every regional area perceive these yearly events as theirs? On the same day as Girl's Day, *hamauri* is held in Okinawa. It is believed to be the day for women to purify themselves in the sea. Women, including girls, are encouraged to go to the beach, pray, and enter the sea in Okinawa (Higa 2008). Nowadays, it has become a custom for women to get together on the beach and enjoy gathering. Morishita (2021) who is based in Okinawa describes people in Okinawa as feeling more attached to the *hamauri* ritual outdoors than to decorating *hina* dolls inside their homes.

Research questions

1. How did autumn leaf viewing become a custom, and why is it still popular?
2. Why do prefectures have their own flowers?

3. What are the symbolic plants of your prefecture/ a prefecture you know in Japan? (or even in your town in your country?)
4. What is the history and significance of the plant in the particular region?
5. Where can you find symbolically represented flowers in the region? (Food produce, Clothing? Advertising? Tourist attractions? etc.)
6. How might the COVID-19 pandemic and 'stay-home' regulations have influenced people's affection towards flowers?
7. Japan seems to have so many kinds of flowers which are symbolically used during each slight change in season. Are there any countries which have similar customs to this?
8. There are lots of names given that are flowers and plants across different cultures. Explore the masculine, feminine, and gender-neutral names that are given in other cultures.

Conclusion

This chapter looked at how flowers, branches, and *ikebana* related arrangements are used for Japanese events and rituals throughout the year and during a person's life. It also examined how these yearly rituals and life events have become simplified in the Japanese household as society has changed. Within these circumstances, it is interesting to see how the related culture such as its relevant flower culture is affected. We learned which flowers were appropriate and which were taboo for specific occasions, as well as the meaning and wishes behind different flower arrangements. This knowledge is significant in order to explore the in-depth meanings of *ikebana* arrangements and flower culture in Japan. With the knowledge we learned from this chapter, let us now examine the aesthetics of these flower arrangements.

References and further reading

Cross, T. (2009) *Ideologies of Japanese Tea: Subjectivity, Transience and National Identity*, Kent: Global Oriental.

Goto, K. (2021) Personal Communication 20 June.

Hartmut, R. (2003) *Social Acceleration A New Theory of Modernity*, Trejo-Mathys (Trans.), Columbia: Columbia University Press.

Hendry, J (2019) *Understanding Japanese Society (5th ed.)*, London: Routledge.

Higa, J. (2008) *Okinawa no Shikitari* [Custom in Okinawa], Tokyo: Sohosha.

Hirabayashi, J. (2015) 'A Buddhist perspective on Japanese spirituality, Japanese spirituality, language, culture, and communication,' *Journal of the College of Intercultural Communication*, 7, 165–181. DOI: 10.14992/00010982

Kaikan, N. (2021) Personal Communication 20 June.

Kawase, T. (2000) *The Book of Ikebana*, Tokyo: Kodansha.

Kitaakita City Board of Education. (2011) *Kitaakitashi Rekishi Bunka Kinon Koso [Kita Akita city History Culture Planning]*, Kitaakita City: Kitaakita City Board of Education. Kojien.

Kobayashi, T. (2019) 'Nihon jin no shukyoteki ishiki ya koudou ha dou kawattaka' [How have Japanese people's religious attitudes and behavior changed?], *ISSP International Comparative Survey on Religion in Japan*, 69(4), 52–72.

Lebra, T. (1995) *Above the Clouds: Status Culture of the Modern Japanese Nobility*, California: University of California Press.

Morishita, C. (2021) Personal Communication 20 June.

Saito, T. (2018) *Nihon no Gyoji* [Japanese Event], Tokyo: Kin no Hoshisha.

Sasaki, H. (2013) 'Namae wo ou' [Chasing the name], *Kazoku shakaigaku kennkyu (The research of families in the society)*, 25(1), 5–6.

Segawa, K. (1982) *Mura no Minzoku* [Village custom]. Tokyo: Iwasaki Bijyutusha.

Shintani, T.(2019) *Kisetsu no Gyoji to Nihon no Shikitari Jiten [Seasonal Event and Japanese Custom]*, Tokyo: Mainabi Bunko.

Shimada, Y. (2010) *Soshiki wa iranai* [We do not need funerals], Tokyo: Gentosha Shinsho.

Shinmura, I. (ed.) (1991) *Kojien*, Tokyo: Iwatani Shoten.

Suzuki, M. (2006) 'Annual events and cuisine in current homes' [gendai-katei ni okeru nennzyuu-gyouzi to tabemono], *Bulletin of Cultural Research Institute, Aoyama Gakuin Women's Junior College*, 14, 3–20.

Yanagida, K. (2019) *Nihon no Minzokugaku* [Japanese folklore], Tokyo: Chuo bunko.

Watanabe, L. (2021) Personal Communication 20 June.

Watanabe, H. (2011) 'Japanese religious population: The answer to the suspicious 20–30% of 2,000,000' [nihon no syu-kyou-zinnkou: 2 oku to 2–3 wari no kai no kai], *Journal of Institute of Buddhist Culture, Musashino University*, 27, 25–37.

2
PHILOSOPHY AND AESTHETICS IN *IKEBANA*

Japanese appear to believe that flowers and branches are beautiful; they feel calm and pleased to look at them. Flowers and branches entice us into an imperfect natural world, but why do we feel this way? This chapter begins exploring the concept of *ikebana* by examining its philosophy and aesthetics: *wabi sabi* and *miyabi*. By sharing examples of these concepts in traditional Japanese arts and culture, it compares the differences between *ikebana* and the aesthetics of western flower arrangements. Readers will also explore how these aesthetics are interpreted outside of Japan and rethink how these are perceived by the younger generation in Japan.

What is *ikebana*? Nature and us

When flowers are arranged in our daily lives, we generally use the word decorate (*kazaru*). In *ikebana* practice, the phrase *ikeru*, literally means to be alive and is used intentionally. *Kazaru* has the connotation that we as human beings are superior to flowers and are using them to decorate in our own way. On the other hand, *ikeru* is interpreted as how practitioners arrange flowers recognizing that they are still alive (Figure 2.1). *Ikebana* has been influenced by Shintoism which believes deities exist in nature and Buddhism philosophy which emphasize that nature is on the same level as human beings: nature, flowers, and plants should be respected (Kawase 2000, Higaki 1989). Freshly cut flowers are understood as still being alive, and thus to show respect for them, practitioners are guided to arrange flowers with both hands. The *ikebana* teacher, Sugidate-*sensei*, often comments, 'Listen to flowers' voices. Feel how they want to be arranged. Arrange

DOI: 10.4324/9781003248682-3

FIGURE 2.1 A practitioner arranging flowers

Source: Photograph by the author.

the flowers by showing your gratitude for letting you arrange them.' At an *ike-bana* exhibition in Akita city at the large building, one of the practitioners said:

> My flowers are so *kawaiso* (poor/pathetic-looking). It's a pity to arrange them here. This room is in a concrete building, so we have powerful air conditioning. My flowers feel so dry and thirsty, they do not last very long. So, I give lots of water to them, as much as I can. Thanks to that, look at them. They are doing their best *(ganbattru)* to show their beauty to the audience. I am so grateful to them. I always tell them to do their best until we finish our exhibition. When we finish the event, I always say *arigato* to them and say goodbye or pass them to someone else.

In *ikebana*, practitioners often have to use the technique of bending a flower's stem at 90 degrees. Sugidate-*sensei* often says, 'Sorry, sorry flower. Please be patient with me a little *(sukoshi gaman shitekudasai)*.' *Ikebana* practitioners seem to treat flowers as if they were people. They talk to them while they arrange flowers as if they talk to people who care. Okakura Tenshin (1863–1913), recognized as the first person to introduce Japanese aesthetics to the West, describes flowers as follows, 'In joy or sadness, flowers are our constant friends. We eat, drink, sing, dance, and flirt with them. We wed and christened with flowers. We dare not die without them' (Okakura [1906] 1989: 109). Okakura further described how flowers were treated in Europe and America, and commented how *ikebana* masters engage with flowers, "He, at least, respects the economy of nature, selects his victims with careful foresight, and after death, does honor to their remains' (Okakura [1906] 1989: 113). These comments from *ikebana* practitioners and Okakura ([1906] 1989) express that one's attitude related to *ikebana* aligns with the philosophy of giving respect to nature and the flowers.

How do *ikebana* practitioners perceive flowers and nature? Emphasizing harmony with nature can be seen from the basic *ikebana* arrangement, triangle shape style. As can be seen from Chart 2.1 the longest branch represents heaven (*ten*),

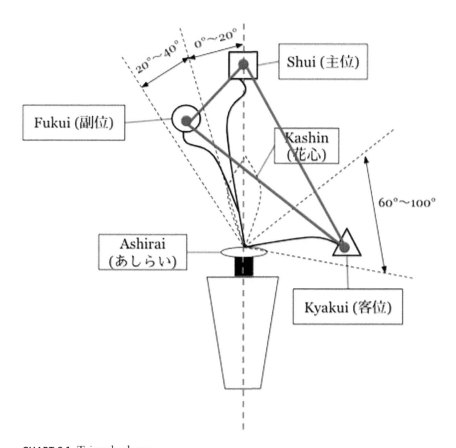

CHART 2.1 Triangle shape

the second and third longest branches represent earth (*chi*), and human (*jin*) respectively, practitioners are encouraged to keep this triangle for emphasizing asymmetrical beauty, but also to learn the harmony between these three elements, heaven and earth, which is interpreted as nature, and the human being (Akitaken Kado Renmei 2007, Sasaoka 2013).

Wabi sabi and *miyabi*

The concept of *wabi sabi* is understood as one of the most well-known Japanese aesthetics. While the definition or the interpretation is slightly different depending on the artist or critic, the fundamental concept is simple: it is the imperfection of beauty (*fukanzenbi*). Originally, *wabi* and *sabi* were two separate concepts that had negative meanings (Handa 2013). *Wabi* described the loneliness of living in nature, far from society, or poorness; *Sabi* described the state of becoming deteriorated over time. However, with the influence of Zen Buddhism which emphasizes the imperfection of life, meditation to be nothingness, the two concepts *wabi* and *sabi* gradually came to be perceived as positive (Juniper 2011). *Wabi* comes to describe the more positive aspects of living alone in nature, such as being modest and being able to appreciate the beauty in the elegance of humbleness and rustic simplicity. The meaning of *sabi* developed into the appreciation of the beauty in old age or old things and focusing on inner beauty that comes from old things. Hisamatsu (2002) describes the seven perspectives of *wabi sabi*: asymmetry, simplicity, wizened austerity, naturalness, elegance, unconditional freedom, calmness, and tranquility. Hisamatsu (2002) also highlights how *wabi sabi* appreciates the cycle of growth, decay, and death, which is the natural order of life for all living things. The spirit of *wabi sabi* can be believed as a very realistic concept, which reinstates how this concept of nothingness is not an issue, as nothing is absolutely ideal and nothing lasts forever.

Miyabi is understood as graceful, sophisticated, elegant, and noble beauty. *Miyabi*, the word itself, is derived from *miya*, which means a loyal court. This concept is derived from the noble culture during the Heian period (Kawase 2000). The concept of *miyabi* detaches from the simpleness and roughness, and keeps the noble engaged with restricted cultural and artistic methods or regulations. Having a tendency to not always be simple and rough, *miyabi* tends to be interpreted as gorgeous (*hanayaka*).

How can we see the *wabi sabi* concept in *ikebana*? As can be seen in Figure 2.2, practitioners are trained to arrange flowers asymmetrically from the beginning, so without thinking they seem not to forget about the emptiness. Some *ikebana* arrangements use withered leaves to express the beauty of transience. In Figure 2.3, for example, the *haran* arrangement uses fresh, matured, and withered leaves to express this idea of impermanence.

FIGURE 2.2 *Ikebana* arrangement

Source: Photograph by the author.

Often, the *miyabi* concept can also be seen in *ikebana* arrangements; the majority of arrangements tend not to be simple, but rather gorgeous (*hanayaka*). This *miyabi* and *hanayakasa* concept in *ikebana* seems to be clearly apparent when it is compared with tea flower arrangements, which are commonly known as typical of a *wabi sabi* arrangement (Figures 2.4 and 2.5).

Ikebana and tea flower

As can be seen from Figure 2.4 above, flowers for *chadō* are more humble-looking, are of a smaller quantity, and the flower container is generally simple. *Ikebana* arrangements, on the other hand, use more flowers and colors, and embrace a more gorgeous appearance. Flower arrangements for *chadō* were influenced by Sen no Rikyū; *chadō* practitioners arrange the flowers so that they mirror how they bloom in the field (Ludwig 1974, Sen 2000). The philosophy of flowers used in *chadō* will be elaborated on further in Chapter 5.

FIGURE 2.3 Arrangement with withered leaves

Source: Photograph by the author.

Dualism

The concept of *miyabi* seems to be different from *wabi sabi*, however, Kawase (2000) argues that this is significantly relevant. Kawase (2000) states that the moment of *miyabi* is greatly appreciated since we understand that *miyabi* does not last forever; he further states that cherry blossoms can also be interpreted as *miyabi*, as well as *wabi sabi*, since the moment of full-bloom is gorgeous, yet it does not last for long. Kawase (2000) further comments that the combination of these opposing aesthetic ideals is related to the Japanese concept of dualism. Japanese culture emphasizes the relationship between opposites, including life and death (*sei shi*), motion and stillness (*dou sei*), and non-daily and daily (*hare ke*), which was described in the previous chapter, and without each other, the beauty of either element alone does not emerge. Figure 2.6 is the *ikebana* work created by Terada *Iemoto* for exhibition in Akita. It expresses life and death. Terada (2021) expresses that without death, there is no appreciation for our current lives.

FIGURE 2.4　Tea flower

Source: Photograph by the author.

Wabi sabi and *miyabi* in other forms of Japanese culture

How can we see these aesthetics in other forms in Japan? The concept of *wabi sabi* is apparent in Japanese interior architecture. It can be seen on the grounds of the Imperial family residence and the rooms into which they welcome delegates from overseas (Abercrombie 2001, Amagasaki 2017). They were invited to the simple rooms surrounded by a refined quality of wood, but less gold, simple decorations with plain furniture. Amagasaki comments (2017: 2) that when these greeting pictures were on the website, there were several comments including, 'The Japanese Emperor needs an interior designer to upgrade their rooms!' The concept of *wabi sabi* is not only indicated by appreciating the emptiness, but also in the communication style. For instance, speaking without any appreciation for a silent moment (*ma*), is not appropriate. People who talk without *ma* are considered impolite or nervous (Kempton 2018, Prusinski 2012).

FIGURE 2.5 *Ikebana*

Source: Photograph by the author.

 Kimonos with detailed colorful decorations, *geisha* and *maiko* performances, and detailed and colorful Japanese lunch boxes (*bento*) are generally understood as *miyabi* (Figures 2.7 and 2.8). The term *miyabi* is often related to graceful yet colorful images of Japanese beauty, and tourist agencies tend to use this term when advertising tours in Kyoto to see cherry blossoms, the colorful changes in autumn leaves, and organizing gatherings with *geisha* and *maiko* in the Gion area. Some food companies have offered detailed *obento* advertising, 'Taste *miyabi* (*Miyabi wo ajiwau*).' The Chinese character for *miyabi* is commonly used for Japanese

FIGURE 2.6 LIFE AND DEATH by Terada *Iemoto*

Source: Photograph by Chikuseikai.

names as parents and family members choose this word wishing their baby to be a graceful, sophisticated, and sociable person. Empress Masako also has this Chinese character in her name.

Flowers are often perceived as either *miyabi* or *wabi sabi*, depending on the occasion (Kawase 2000). *Miyabi* descriptions seem to be used for fully bloomed Japanese apricot flowers, such as those in the Kairakuen garden in Ibaraki Prefecture. This Japanese garden was owned by the Tokugawa family and over 3,000 Japanese plum trees bloom around the end of February. On the other hand, the butterbur (*fukinoto*), just sprouting from the ground from underneath the snow can be expressed as *wabi sabi*. Rikyū felt the *wabi sabi* aesthetic to resonate with the following poem by Fujiwara no Ietaka (1158–1237);

> To those who await
> Only the cherry blossoms
> One should show the grasses of spring
> Emerging through the snow
> In a mountain village
>
> *Sen 1989: 239*

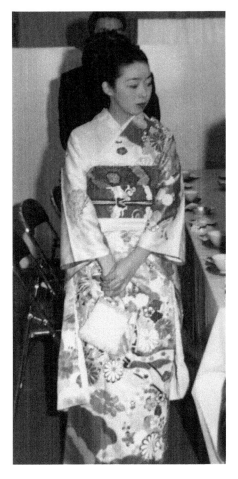

FIGURE 2.7 *Kimono*

Source: Photograph by Minoru Yamada.

Western flower arrangement

How are the aesthetics of *ikebana* different from Western flower arrangement? Western flower arrangement places an emphasis on perfectly symmetric shapes to express the beauty of manpower (Figure 2.9). Western flower arrangement is generally filled with flowers with no gap. To study the beauty of *ikebana*, practitioners are also encouraged to take western flower arrangement lessons as the part of their curriculum, Sugidate-*sensei* commented on the lesson:

> Western flower arrangement is so different from *ikebana*. It seeks for the perfectly dormy shape, no emptiness, the flower arrangement should be identical from left to right. We do not use branches like *ikebana*, most of

FIGURE 2.8 *Akita Maiko*

Source: © Akita Maiko.

them are flowers. First, we place the biggest flower on the center of the oasis, then place the second biggest flower on the right side of the corner, then find the identical flower, and place that one to the left side. The Western one appreciates beauty by addition, but *ikebana* is the beauty of subtraction. *Ikebana* uses fewer flowers, it removes unnecessary elements and it expresses the flower's life from blooming to withering.

FIGURE 2.9 Western flower arrangement

Source: Photograph by the author.

Sugidate-*sensei* significantly emphasizes the symmetric beauty of western flower arrangement. A flower itself is never the same from one to another, and it is not easy to find an identical one. On the other hand, *ikebana* tends to emphasize the individual differences, or individual nature (Kawase 2000). *Ikebana* aesthetics are influenced by Buddhist values holding that humans and nature are equally important. This means *ikebana* is stylized as naturally as possible: An *ikebana* arrangement is asymmetrical to express the natural beauty, and it generally emphasizes the emptiness (*ma*).

Similar differences can be seen in gardens: Western gardens are generally symmetric, the trees are trimmed in an artificial way to express the artificial beauty, and fountains are often used to show manpower by changing the direction of the water flow from the ground up. In contrast, the water in a Japanese garden is stylized as naturally as possible, with small waterfalls above modest ponds. The way the garden is designed generally aligns with the image of nature in Japan; the landscape style includes using small mountains or hills, waterfalls, rocks, and pathways (Naka 2009).

Wabi sabi outside of Japan

The *wabi sabi* aesthetic also influences artists outside of Japan. Claude Monnet (1840–1926) is one of the greatest examples. He was impressed with Japanese paintings, including Katsushika Hokusai (1760–1849), and therefore applied the *wabi sabi* aesthetic to his painting (Handa 2013, Hiramatsu 2016). As described above, Okakura Tenshin (1863–1913) is recognized as the first person to introduce Japanese aesthetics to the West. Okakura's book, *Book of Tea*, was published in 1906 in Boston, USA. In his book, he stated, 'For Teaism is the art of concealing beauty that you may discover, of suggesting what you dare not reveal' (Okakura ([1906] 1989: 7). While the term *wabi sabi* was not used by exploring *chadō* and flower culture in Japan, Okakura described the difference between western and eastern aesthetics. He explained that our imperfect lives themselves are Teaism (Yixuan and Song 2021).

According to the current preferred aesthetics survey in Japan, around 70 percent of people find beauty in 'rational, functional and practical' objects, 60 percent of people in 'simple, discreet, unclear, and imaginable' things, 30 percent of people in 'flashy, gorgeous, and glamorous' items (Ministry of Land, Infrastructure, Transport, and Tourism 2019). How do we interpret this result? Nowadays, these *wabi sabi* aesthetics and ideology seems to be appreciated overseas even more than in Japan. *Wabi sabi* related books have been published more in English than in Japanese since 2015. Related to the major influence of social media, people appear to be facing an ideal 'perfect life,' which has become a stressful life (Kempton 2018). In these circumstances, there seems to be a tendency to embrace *wabi sabi* aesthetics and ideology in relation to mindset, health, and interior trends. For instance, KonMari, a Japanese consultant guiding the simple lifestyle, has become well supported in North America. In terms of interior designs, simple handmade decorations are appreciated more than matched cohesive designs. On the other hand, *wabi sabi* aesthetics by overseas artists re-enter Japanese society with new and interesting interpretations. Tom Sachs, an artist from New York expresses in-depth beauty and the philosophy of *wabi sabi* with the space images. His exhibition 'TEA CEREMONY' opened in Japan and Japanese media outlets introduced his artwork as stimulating and exciting (Matsumoto 2019).

Research questions

1. Is this concept of having a particular aesthetic, such as *wabi sabi* and *miyabi*, only a Japanese thing?
2. Which description of aesthetics from Okakura do you find relevant to your life?
3. *Wabi sabi* may have contributed to the Japanese respecting silence, which nowadays may not be perceived as a positive attitude. As people become more expressive of their feelings, will the idea of *wabi sabi* be less significant?

4. How has the concept of *wabi sabi* been transformed outside of Japan?
5. Why does the incompleteness of *wabi sabi* feel comfortable in our lives?
6. Are *wabi sabi* and *miyabi* concepts the obstacle for promoting *ikebana* in the future?
7. Can the *wabi sabi* concept apply to current popular culture?

Conclusion

This chapter explores the philosophy and aesthetic of *ikebana*. By sharing examples of *miyabi* and *wabi sabi* in traditional Japanese arts and cultures, this chapter has a close look at how these aesthetics can be seen in other art or cultural forms across and outside of Japan. It is also interesting to examine how the younger generation perceives these aesthetics. Cross (2009) argues that the younger generation in Japan is no longer interested in traditional arts due to technological development, as every moment in our lives requires efficiency and quick responses – even in our social relationships. The lifelong training style is no longer preferred, but does that necessarily mean that *wabi sabi* concepts are no longer appreciated? The term '*emoi*' used among the younger generation appears to be used when they are also commenting on matters of transience. Not possessing cars, brand bags, and clothes but instead renting them, some may argue that this lifestyle of the young generation is similar to Rikyū's philosophy of *wabi sabi* life.

References and further reading

Abercrombie, S. (2001) 'Wabi Sabi: A new look at Japanese design / wabi sabi style,' *Interior Design*, 72(8), 125.

Akitaken Kado Renmei (2007) *Akitaken Ikebana shi* [Akita Ikebana History], Akita: Akita Kyodo Press.

Amagasaki, A. (2017) 'What can be seen and what cannot be seen- the aesthetic sense of Japanese medieval arts,' *The Gakushuin Journal of International Studies*, 4, 1–14.

Ankermann, P. (1997) 'The four seasons: one of Japanese culture's most central concepts,' *Nordic Institute of Asian Studies Man and Nature in Asia* (1), 36–53.

Cross, T. (2009) *The Ideologies of Japanese Tea: Subjectivity. Transience and National Identity*, Kent: Global Oriental.

Graham, P. (2014) *Japanese Design: Art, Aesthetics & Culture*, Boston: Tuttle Publishing.

Haga, K. (1989) 'The wabi aesthetic through the ages,' in Kumakura and Varley (eds.), *Tea in Japan: Essays on the History of Chanoyu*, Honolulu: University of Hawaii Press, pp. 195–232.

Handa, R. (2013) 'Sen no Rikyū and the Japanese way of tea: Ethics and aesthetics of the Everyday,' *Interiors*, 4(3), 229–247.

Hashikawa, B. (1982) *Okakura Tenshin Hito to Shiso* [Okakura Tenshin, People and Philosophy], Tokyo: Heibonsha.

Higaki, T. (1989) 'Japanese animism and the religious view of nature: The discovery of a special Japanese religious consciousness,' *Journal of Esoteric Buddhism*, 1989(165), 1–31. DOI: 10.11168/jeb1947.1989.165_1

Hiramatsu, R. (2016) *Mone to Japonism* [Monet and Japonisme], Tokyo: PHP Institute.

Hisamatsu, S. (2002) *Zen Talks on The Record of Linji. Hisamatsu's Talks on Linji*, Hawaii: University of Hawaii Press.

Iwai, S. (2006). '"Nihonteki" biteki gainen no seiritsu (2): Chado wa itsukara "Wabi" "Sabi" ni nattanoka?' [The establishment of 'Japanese' aesthetical concept (2): When did the tea ceremony started performed under the concept of 'Wabi' 'Sabi'?], *Nihon Kenkyu* [Japanese Study], 33, 29–53.

Izutsu, T. and Toyo, I. (1981) *The Theory of Beauty in the Classical Aesthetics of Japan*, Springer Science & Business Media.

Juniper, A. (2003 [2011]) *Wabi Sabi: The Japanese Art of Impermanence*, Boston: Tuttle Publishing.

Kawase, T. (2000) *Book of Ikebana*, Tokyo: Kodansha International.

Kempton, B. (2018) *Wabi Sabi: Japanese Wisdom for a Perfectly Imperfect Life*, London: Piatkus Books.

Koren, L. ([1994] 2008) *Wabi-Sabi for Artists, Designers, Poets & Philosophers*, Point Reyes: Imperfect Publishing.

Kubota, R. (2003) 'Critical teaching of Japanese culture,' *Japanese Language and Literature*, 37(1), 67–87. DOI: 10.2307/3594876

Kurita, I. (2001) *Traveling with Flowers*, Tokyo: Iwanami Shoten.

Ludwig, T. (1974) 'The way of tea: A religio-aesthetic mode of life,' *History of Religions*, 14(1), 28–50.

Matsumoto, M. (2019) 'The museum becomes a "tea ceremony space." Contemporary artist Tom Sachs' extraordinary tea ceremony,' *The New York Times Style Magazine Japan*, 16 May. www.tjapan.jp/art/17267144 (accessed 14 October 2021).

Matsuo, Y. (2014) 'Wabi-sabi to chado' [Wabi-sabi and the tea ceremony], *Graduate School of Human Ecology, Kinjo Gakuin University*, 14, 61.

Ministry of Land, Infrastructure, Transport and Tourism. (2019) The White paper on Land, Infrastructure, Transport and Tourism in Japan, 2019: Part I Chapter 1 Section 3. www.mlit.go.jp/hakusyo/mlit/h30/hakusho/r01/pdf/np101300.pdf (accessed 14 October 2021).

Mitsui, H. (2008) *Katachi no Nihonbi: Wa no designgaku* [Japanese beauty in shape: Design studies of 'wa.'], Tokyo: NHK Shuppan.

Molin, S. (2020) *Wabi Sabi: The Philosophy of achieving perfection through Imperfection*, independently published.

Naka, T.(2009) *Nihon Teien no Mikata* [The way to look at Japanese Garden], Tokyo: Tokyo Bijyutsu.

Okakura, K. (1904) *The Ideals of the East with Special Reference to the Art of Japan*, New York: E.P. Dutton & Co.

——— ([1906] 1989) *The Book of Tea*, Tokyo: Kodansha International.

Okakura, K. and Dalby, L. (2000) *The Book of Tea: Beauty, Simplicity and the Zen Aesthetic*, Boston: Tuttle Publishing

Oooka, M. (1985) *Okakura Tenshin*, Tokyo: Asahi Shimbun.

Prusinski, L. (2012) 'Wabi-sabi, mono no aware, and ma: Tracing traditional Japanese aesthetics through Japanese history,' *Studies on Asia*, IV, 2(1), 25–49.

Saito, Y. (1997) 'The Japanese aesthetics of imperfection and insufficiency,' *The Journal of Aesthetics and Art Criticism*, 55(4), 377–385. DOI: 10.2307/430925

Sasaoka, R. (2013) *Ikebana*. 37(3), 181–186. DOI: 10.11469/koshohin.37.181

Sen, S. (1979) *Tea Life, Tea Mind*, trans. Urasenke Foreign Affairs Section, Urasenke Foundation, New York: Weatherhill.

——— (1989) 'Reflections on Chanoyu,' in Kumakura, I., and Varley, H. P. (eds.), *Tea in Japan: Essays on the History of Chanoyu*, Honolulu: University of Hawaii Press, pp. 233–254.

———— (2000) *Urasenke Chado Textbook*, trans. Urasenke Foreign Affairs Section, Kyoto: Urasenke Tankokai.

Shirane, H. (2017) 'Japan and the culture of the four seasons: Nature, literature and the arts, rethinking nature in contemporary Japan: Tradition to modernity,' *Ca' Foscari Japanese Studies*, 7, 79–26. DOI: 10.14277/6969-171-3

Suzuki, D. T. ([1938] 1970) *Zen and Japanese Culture*, Princeton: Princeton University Press.

Takashina, S. (2015) *Nihonjin nitotte Utukushisa toha Nanika* [What is the beauty for Japanese?], Tokyo: Chikuma shobo.

Tai, E. (2003). 'Rethinking culture, national culture, and Japanese culture,' *Japanese Language and Literature*, 37(1), 1–26. DOI: 10.2307/3594873

Terada, K. (2021) Personal Communication 1 July.

Yixuan, S., and Song, J. (2021) 'The influence of cultural information design and the classification of spiritual guidance, based on the cases of Japanese wabi-sabi,' *International Journal of Contents*, 17(1), 27–36.

Yoshimura, K. and Yamada, Y. (2017) 'Wabi and sabi as the background of Japanese color aesthetics: Pursuing the special qualities of Japan's traditional color sensibility,' *Japan Color Bulletin*, 41(3), 40–43.

Movies

Kumai, K. (1989) *Sen no Rikyū*. [Death of a Tea Master], Japan: Toho.

Onori, T. (2018) *Nichi Nichi Kore Kojitsu* [EveryDay A GoodDay], Japan: Tokyo Theatres.

Ooji, M., Kazutaka, F., Hidetomo, S. (Producer), & Mitsutoshi, T. (Director). (2013). *Rikyū ni Tazuneyo*. [Ask This of Rikyū]. Japan: Toei.

Shinohara, T. (2017) *Hanaikusa*. [Flower and Sword]. Japan: Toei.

Teshigahara, H. (1989) *Rikyū*, Japan: Shochiku.

3
HISTORY

Ikebana history introduces how the flower culture and *ikebana* have evolved throughout time (Chart 3.1). It explores how flower arranging in Japan started with a religious influence and gradually changed alongside architectural styles. The reader will examine how the aesthetics, politics, economy and lifestyle of each era have all played roles in the evolution of Japanese flower culture. It further traces how *ikebana* culture has evolved in Akita with some interviews from *ikebana sensei*. There is a lot to learn about the history of *ikebana*; examining the topic closely will reveal how *ikebana* culture and practice can be revived in contemporary times.

From antiquity through the Asuka period (592–710): the influence of Animism

The ancient Japanese appear to have believed that there were deities in natural elements of the world, such as mountains, oceans, rocks, wind, and flowers; this is also known as Animism. It was a common belief that deities did not have a permanent place to stay and therefore constantly moved around. Tall evergreen trees were thought to be a kind of antenna for deities to find a place to stay (Figure 3.1). These trees are called *yorishiro* (Akitaken Kado Renmei 2007). Kawase (2000) commented that the custom of worshiping these trees has influenced *ikebana* style and other flower-related decorations. The classical *ikebana* arrangement, *rikka* style, uses tall evergreen branches with other flowers and materials (Figure 3.2). As described in Chapter 1, one kind of New Year decoration, the *kadomatsu* at the entrance hall, is also believed to hold an antenna-like function for the god of the New Year, *toshigami-sama*.

DOI: 10.4324/9781003248682-4

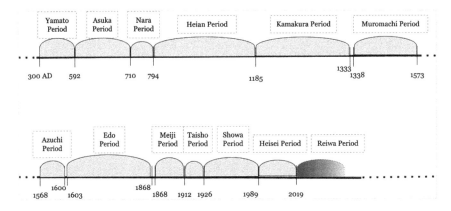

CHART 3.1 Japanese history chart

FIGURE 3.1 *Yorishiro*

Source: Photograph by the author.

FIGURE 3.2 *Rikka*

Source: Photograph by the author.

The Nara period (710–794): influence from Buddhism

Buddhism originated in India, flourished in China and then spread among nobles in Japan, thereafter heavily influencing the flower arrangement practices of the time. Though Shintoism holds that deities are homeless, Buddhism believes that the Buddhas have tangible images that should be housed in special sacred structures or temples (Herley 1990). Thus, the first act of *ikebana* was when monks created ritual flower offerings (*kuge*) at Buddhist altars. This custom of arranging flowers in the Buddhist context can still be seen; flowers, especially chrysanthemums (*kiku*), are often arranged at family altars within private homes.

The Heian period (794–1185): the custom of admiring flowers is born

In the Heian period, the Fujiwara family held power and noble culture flourished. Political and formal diplomatic relations between China and Japan, meanwhile, eventually ceased. Trade and cultural exchange, including goods,

Chinese literature, writing, and customs became sporadic. Japan, in short, was much less culturally influenced by China, and as a result, a unique Japanese culture developed. Buddhist customs also evolved into a Japanese style, including the way of arranging flowers. Originally, only the petals or the crowns of flowers would be arranged at the Buddhist altar, but the whole flower including the stem started to be used during the Heian period. Kawase (2000) states that this style was influenced by the *yorishiro* arrangement, which became common among aristocrats. *The Tale of Genji* (around 1021) and *The Pillow Book* (1002) were written by court ladies, Murasaki Shikibu (973 or 978–1014 or 1031) and Sei Shōnagon (966–1017 or 1025). Murasaki Shikibu described all the flowers in the tall vase in front of the Buddhist scroll as one beautiful scene in the courthouse (Akitaken Kado Renmei 2007, Mizue 1985). Kawase (2000) argues that this arrangement has influenced the Tatehana style in *ikebana*. He further stated that this style encouraged nobles in the Heian period to admire the beauty of flowers themselves.

Influenced by the architecture of Buddhist temples, the *Shinden-zukuri* architectural style was developed and became popular among aristocrats (Figure 3.3). This style emphasized the indoor lifestyle. Within their mansions, aristocrats would plant trees and plants in inner courtyards, and the custom of cherishing flowers emerged (Imai 2000). There are numerous poems written by Murasaki Shikibu and Sei Shōnagon describing the beautiful flowers which could be seen from inside the house. Many such poems are still admired today, notably the following: *Kokoroateni sorekatozomiru shirotsuyuno hikarisoetaru yuugao no hana*, Are you Genji? You are just like the *yugao* shining with the dew drop (Takishima 2006: 365).

The Azuchi-Momoyama period (1568–1600): Sen no Rikyū and *nageire*

Under the rule of warlords Oda Nobunaga and Toyotomi Hideyoshi, *chadō* became a well-known culture and taste among the warrior and merchant classes. Sen no Rikyū, the founder of *chadō*, was also famous for promoting tea-related flower arrangements for tea gatherings. Compared to the *rikka* style, this arrangement is natural and simple. *Ikebana* was also influenced by Sen no Rikyū's aesthetics and a similar style of *chadō* arrangement, which introduced the nageire style into *ikebana* (Akitaken Kado Remmei 2007).

The Edo period (1600–1868): the *iemoto* system emerges

The Tokugawa family ruled Japan hegemonically for over 260 years, during which time the government restricted trade and limited contact with the wider world. This arguably resulted in a golden age for traditional Japanese arts and

FIGURE 3.3 *Shoin zukuri*

Source: Photograph by the author.

culture, including *ikebana*, *chadō*, the way of incense (*kodō*), and Japanese trad-itional dance (*nihon buyō*) (Okakura [1906] 1989). These forms of traditional arts became widely practiced by the merchant class, and the number of practitioners increased dramatically. In order to maintain a sense of order among the growing number of practitioners, the *iemoto* system emerged (Imai 2000). In reference to flower arranging, the *rikka* style was regarded as too large, time-consuming, and requiring intense skill at arrangement. In order to accommodate the needs of the merchant class, the Ikenobo School created the *shōka* or *seika* style which created simpler arrangements. This style only has three main branches, which form a unity expressing life's perpetual change and renewal (Figure 3.4). While *ikebana* and *chadō* were mainly for men, upper or upper-middle class women started to engage with these art forms in their private lessons from the middle of the Edo period (Akitaken Kado Renmei 2007, Sasaoka 2013) (Figure 3.5). These customs led to *ikebana* education being introduced in girls' schools from the Meiji period onwards.

FIGURE 3.4 *Saika*

Source: Photograph by the author.

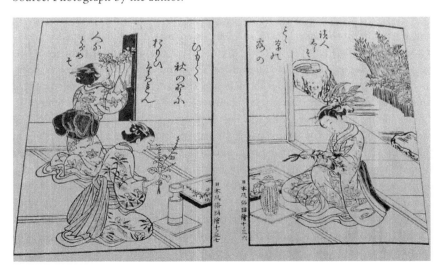

FIGURE 3.5 Women practicing *ikebana* in the Edo period

Source: © Akita Prefectural Library.

The Meiji period (1868–1912): *ikebana* flourishes despite the challenge of westernization

The Meiji revolution brought a financial crisis to traditional arts in general, but especially to *noh* and *chadō*, which had substantial support from the leading *samurai* class. On the other hand, *ikebana* recovered from their crisis quickly since they had a lot of financial support from non-leading *samurai* classes including scholars and future Meiji government officers (Corbett 2018).

The Meiji government strived to promote Japan as westernized; the aim was to enrich the country and strengthen its military, and thereby become equally as powerful and significant as western countries (Corbett 2018). Due to westernization policies, traditional arts including *ikebana* and *chadō*, which were thought to hold feudalistic sentiments, were not promoted by the government at the beginning of the Meiji period. However, *ikebana* was regarded as a valuable factor in raising a young woman to be 'a good wife and wise mother (*ryōsai kenbo*)' and was selected as an official subject for girls' schools (Corbett 2018, Ito 2016, Shimizu 1996). Starting in the Meiji period and ever since, *ikebana* has been a common subject in girls' schools. One of the leaders from the Ikenobo School, Mutō, stated that *ikebana* philosophy emphasized mental training, not just learning of flower-arranging techniques per se. Akitaken Kado Renmei (2007) argues that this philosophy was highlighted in the mass media and consequently gained wide support in Japanese society. Although as part of the curriculum at girls' schools it might appear to have become fully mainstream, *ikebana* was in practice restricted to upper class women since at the time only wealthy families could afford to send their daughters to girls' finishing schools, which were secondary schools and as such over and above mandatory schooling. Additionally, there were only male *ikebana* teachers until the end of the Meiji period due to the idea of male superiority (Akitaken Kado Renmei 2007). This gender issue is explored more in the gender and class chapter.

Numerous flowers from the West were imported and attracted the attention of the Japanese. Unshin Ohara (1861–1916) created the *moribana ikebana* style. Compared to *rikka* and *saika* styles, *moribana* only required a simple technique and used a variety of western flowers (Figure 3.6). The *moribana* style, which encouraged using a variety of colors of western flowers, became well accepted in Japan, particularly among young women, once Unshin established the Ohara school in 1895 (Akitaken Kado Remmei 2007).

The Taisho (1912–1926) and early (1926–1945) Showa periods: *ikebana* golden age

Throughout this era, democracy and its philosophies grew in number and strength throughout Japan, unsurprisingly also influencing *ikebana* styles at the same time. Democracy was widely spread in Japan, and the *ikebana* style seemed to be influenced by this philosophy. Teshigahara Sōfū (1900–1979) applied the

FIGURE 3.6 *Moribana*

Source: Photograph by the author.

western concept of art to *ikebana* and established the Sogetsu school in 1925. He perceived plants as sculptural elements and created a freestyle approach toward *ikebana* (Imai 2000). Freestyle is now well accepted by other *ikebana* schools and has become an attractive characteristic of *ikebana*. Around this time, *ikebana* spread beyond only upper class women to those in the upper-middle class as well. Other *ikebana* schools also started to be established and attract practitioners. One of these is the Sensho Ikenobo School which was established by Moroizumi Yudo (1881–1952). He had learned *ikebana* in an Ikenobo school at a young age, but it wasn't until his later years that he founded his own school (Akitaken Kado Renmei 2007). This school also pursued etiquette and manners training along with *ikebana* teaching. The current grandmaster promotes the use of British flower arrangement styles for *ikebana* practice.

The postwar period (1945–present): *ikebana* in flux

Numerous policies, voting and property laws, and education requirements were questioned and changed after the Second World War in Japan. Nine years of education became compulsory and *ikebana* was taught mainly as an extra-curricular activity (*bukatsudo*) from the 1950s (Akitaken Kado Renmei 2007). In societal efforts to sustain the Japanese image of the 'good wife, wise mother' *ikebana* was considered an ideal course of study for young women's bridal training (*hanayome shūgyō*). As *ikebana* became affordable to middle-class women, the number of practitioners continued to increase up until the 1970s. Unlike other traditional arts such as *chadō* and *noh*, many *ikebana* schools opened, and now there are over 300 different *ikebana* schools.

Ikebana also spread overseas after the Second World War. During the US Occupation, *ikebana* became popular among SCAP General Headquarters (GHQ) officers' wives. They used *ikebana* lessons as social gatherings in Japan, and many of them continued to practice or engaged in *ikebana* after they went back home. From 1956 to 1959, twenty overseas branches were opened, and the *Ikebana* World Convention had its first meeting in 1965. The word *IKEBANA* became an internationally recognized word (Akitaken Kado Renmei 2007).

Gradually, however, women's roles in Japanese society shifted and bridal training became less popular among women. As a result, the number of practitioners has been decreasing drastically since the beginning of the new millennium. According to Kudo (1995), the number of *ikebana* practitioners was around 30,000,000 in the 1970s, but had fallen to around 10,000,000 in the 1990s (Imai 2000). Muzutani (2020) and Hamazaki (2021) further analyzed the national survey and stated that there were now only around 2,000,000 *Ikebana* practitioners in 2016 (Graph, 3.1, Chart 3.2).

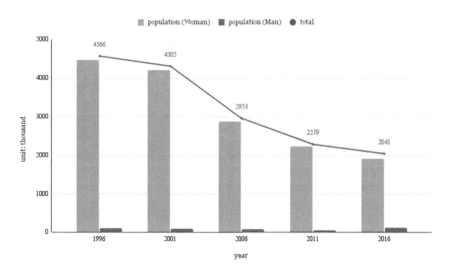

GRAPH 3.1 Change in the population (total and male/female) who practice flower arrangement as a hobby

Source: Ministry of Internal Affairs and Communications 2016.

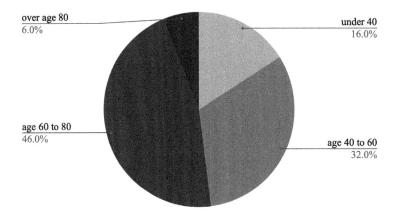

over age 80
6.0%

under 40
16.0%

age 60 to 80
46.0%

age 40 to 60
32.0%

CHART 3.2 Percentage of flower arrangement population by age group in 2016

Source: Hamasaki 2021.

Gorgeous *ikebana* arrangements such as the *moribana*-style arrangements were popular during times of economic expansion. By contrast, a simpler and more humble-looking arrangement, *hitoeda hitohana*, was created by the Sensho Ikenobo *iemoto* after the recession (Sugita 2021) (Figure 3.7). Sugita-*sensei* comments,

> Oh yes, you may not be able to imagine in the 1990s when the economy was bubbly, indeed the *ikebana* was so gorgeous. Practitioners loved using many different kinds of flowers. But now, we rarely see it. Our *iemoto* showed the humble-looking *ikebana*. For this style you only need two different kinds of materials. This is also suitable for our current accommodation situation in Japan; a small condominium in Tokyo does not have the space to arrange a big *moribana* style, right? People just need a small *ikebana* arrangement in the entrance hall. I personally think this is such a great style for our lifestyle.

Ikebana history in Akita prefecture

So far, we have examined *ikebana* history in Japan more generally. As one might assume, Ikenobo started in Kyoto, and the other two largest schools, Sogetsu and Ohara, were established in Tokyo areas. What about a more rural location like Akita? When did *ikebana* start in Akita?

Senke Seika style *ikebana* was introduced and spread in Akita city from the middle of the Edo period. *Senke Seika* was founded by a descendant of Sen no Rikyū, Sen Sogo, who lived in the Yuzawa area in Akita prefecture for a couple of years. During his stay, he taught *Senke Seika ikebana*, and this style is still carried on in Akita today (Akitaken Kado Renmei 2007). By the end of the

FIGURE 3.7 *Hitoeda* style

Source: Photograph by the author.

Edo period, the Enshu School was introduced by Taihei who learned it in Edo (Tokyo). Ikenobo, Senke, and Enshu were the well-practiced schools in Akita during the Meiji period. While practitioners got together and practiced *ikebana*, it became customary to show *ikebana* in the center of Akita by the 1920s. The Sakigake Newspaper article of September 25th, 1920, stated that local *ikebana* teachers tried to minimize the cost of *ikebana* lessons. Akitaken Kado Renmei (2007) argued that *ikebana* targeting a wider group also started around this time in Akita.

Ikebana also became a subject at girls' schools and was taught for the first time in Akita at Akita Kita High School for Girls in 1906 (Figure 3.8). Senke style *ikebana* was taught by this school until 1943. *Moribana* arrangements also became very popular in Akita. A vice-*iemoto* of the Toko Koryu *ikebana* school, Sugimura returned to Akita and taught the *moribana* style, which became very popular. There was some criticism directed toward *moribana* from *ikebana* teachers who preferred the classical *ikebana* style in Akita; however, *moribana's* popularity greatly overshadowed such criticism (Akitaken Kado Renmei 2007).

FIGURE 3.8 *Ikebana* class in 1906

Source: Photograph by a person unknown.

After the Second World War, all the different *ikebana* schools became members of the Akita *Ikebana* Association; they started to have *ikebana* exhibitions and seminars together. In 1957, they offered exhibitions with 12 different schools. One of the biggest *ikebana* schools, Chikuseikai, was established around this time in Akita city. Terada *iemoto*, the head of Chikuseikai, shared her Chikuseikai stories:

> My father was working in the municipality of Akita when he decided to open his school, Chikuseikai. In the end, he quit his job and became *ikebana iemoto*. Interesting, right? He had gone to an agricultural high school in Akita, where he became interested in flowers, and ever since, he had wanted to study *ikebana*. Eventually, he started to learn it from Ikenobo and Sogetsu school teachers in Akita. He really liked it and decided to become an *ikebana* teacher. Fortunately, he had a lot of students, I assume thanks to the bridal training boom. He also opened the Akita *Ikebana* Association and united the *ikebana* schools together. I heard that they did not get on well at all at the beginning; each of the *iemoto* thought that he alone should be head of the association, so in the end, my father asked the mayor to be the head. This shows my father was indeed clever and very good at politics.

Terada *iemoto* continued:

> But in the end, we were working well together. We often organized seminars, inviting some great artists to stimulate our creativity. We held

discussion seminars to think about future *ikebana* design. We still continue this study seminar together; the next one will be held this coming August.

At its peak, there were twenty different schools in Akita prefecture that actively practiced *ikebana*. However, due to the similar issues affecting other prefectures, the number of practitioners has decreased dramatically and there are very few people from the younger generations currently interested in *ikebana* in Akita. Sugita-*sensei* commented during her interview:

> I think this is really the critical time. The practitioners nowadays who are actively engaged are over 70. They will probably continue to engage with *ikebana* for 10 more years, but after that...? Who will do *ikebana*? Who will be the head of this *ikebana* association in Akita and pass it down to children?

Other *ikebana sensei*, including Terada *iemoto*, are also worried about this issue; she comments:

> As *iemoto*, I have tried many promotions to attract many different groups of people. I organized trial sessions at exhibitions. Of course, it is fun and great to organize sessions where beginners could try their hand at *ikebana* hold, but no one mentioned that they wanted to practice further.

It appears that most teachers are well aware of this issue and are doing what they can to attract younger practitioners. They also commented that there is not enough financial support from the municipality, and they seem to struggle promoting *ikebana* culture by themselves.

Research questions

1. Why did *ikebana* not become widespread and popular among women until the Edo era despite the fact that many noblewomen admired flowers starting in the Heian period?
2. The importation of foreign flowers created more outlets for aesthetic expression in *ikebana*. What other ways have practitioners found to express themselves?
3. If *ikebana* had not been picked as a girls' finishing school subject in the Meiji period, do you think *ikebana* today would be more or less popular than it currently is?
4. Why was *ikebana* connected with the idea of 'good wife wise mother' even though at the time almost all the major figures in the art were male?
5. *Ikebana* used to be a part of young women's bridal training. However, as seen with the now-popular buzzword/concept '*ikumen*,' some men nowadays

seem to enjoy child-rearing and doing housework. Since these domestic tasks were once viewed as primarily the role of women in the home, will this trend inspire more men to become active in the practice of *ikebana*?

Conclusion

This chapter explored how flower culture and *ikebana* have evolved throughout each era in Japanese history. *Ikebana* has been constantly evolving, uncovering new styles and techniques to attract new audiences and to reflect the ever-changing social contexts of society in Japan and around the world. Freestyle *ikebana* was a provocative new development to traditional Japanese arts, which tended to emphasize following the strict style (*kata*). However, the Ikenobo School, the oldest and largest *ikebana* school, also accepted the freestyle approach, and this style is often used even at *ikebana* exhibitions featuring classical arrangements. This might be one of the reasons that *ikebana* culture never died out. Just as no two flowers are quite the same, neither are any two *ikebana* arrangements, even if the same flower is used in both. The privilege of freedom and expression allowed *ikebana* not only to survive, but to evolve and prosper, delighting those who appreciate it. That being said, the number of practitioners has decreased significantly in Japan. To secure its future, *ikebana* may need to shift its target and image once more, just as it did in the Meiji period. The topic of how best to promote traditional arts to attract younger generations is covered in the education chapter and the Conclusion.

References and further reading

Akitaken Kado Renmei (2007) *Akitaken Ikebanashi* [Akita Ikebana History], Akita: Akita Kyodo Press.

Corbett, R. (2018) *Cultivating Femininity: Women and Tea Culture in Edo and Meiji Japan*, Honolulu: University of Hawai'i Press.

Hamazaki, E. (2021) 'Kado no genjyo to kadai ni taisu ikebana ryoho gainen no kakushinsei ni kansuru jitusen teki kousatsu [A study of ikebana therapy through investigating contemporary issues in ikebana],' *Institute for the Study of Humanities and Social Sciences, Doshisha University. Shakai Kagaku*, 50(4), 147–175.

Harvey, P. (1990) *An Introduction to Buddhism. Teaching, History and Practices*, Cambridge: Cambridge University Press.

Imai, T. (2000) 'Ikebana-its stagnant situation and the facts study 1: A fact-finding on the history of ikebana,' *Journal of Kyoto Seika University*, 18, 107–129.

Ito, Y. (2017) 'Meijiki iko no taishu ni okeru kado to jenda ni tsuite: joshi kyoiku no shiten kara' [Kado and gender in the mass since the Meiji Period: from the perspective of girls' education], *Waseda Studies in Social Sciences. Extra Issue, Students Journal in 2016 50-year Anniversary Edition for the Establishment of School of Social Science*, 187–196.

Kawase, T. (2000) *The Book of Ikebana*, Tokyo, Japan: Kodansha International.

Kudo, M. (1995) *Nihon Ikebana Bunkashi* [Ikebana History in Japan], Tokyo: Dohosha.

Ministry of Internal Affairs and Communications, Statistics Bureau (2016) *Social Life Basic Survey. Economic census for business activity. 2016 economic census for business activity*, Retrieved from www.e-stat.go.jp/dbview?sid=0003215460 (accessed 8 September 2021).

Mizue, R. (1985) 'Ikebana no rekishi' [History of the flower arrangement in Japan], *Seikatsu Kagaku Kenkyu Seikatsu Kagaku Kenkyu, Bulletin of Living Science*, 7, 98–101.

Mizutani, T. (2020) 'Ikebana sono gendai to koten [Classic ikebana and modern ikebana], *Ikawa Senshinkan Journal*, 5, 1–9.

Okakura, K. ([1906] 1989). *The Book of Tea*, Tokyo: Kodansha International.

Sasaoka, R. (2013) *Ikebana*. 37(3), 181–186. DOI: 10.11469/koshohin.37.181

Shimizu, Y. (1996) 'Nisshin, nichiroki no ie ishiki' [The concept of ie during the Nisshin war and Nichiro war], *Journal of History, Hosei University*, 48, 105–119.

Sugita, M. (2021) Personal Communication 9 July.

Takashima, K. (2006) *A Study of Plants in the Tale of Genji (Vol. 1)* [Genji Monogatari Shokubutsu kou (1)], Kokken Shousho.

Terada, K. (2021) Personal Communication 7 July.

4

IEMOTO SYSTEM

As described in the previous chapter, the *iemoto* system was developed as a way to deal with the popularization of *ikebana*, mainly from the Edo period (1603–1868). This *iemoto* system helped the *iemoto* him/herself to manage a large number of practitioners. Most of the traditional Japanese arts are still maintained through this system. This chapter explores the characteristics of this system, including its licensing (*kyojō*) and regulations. This chapter also introduces the different schools in the *ikebana* field and other traditional arts including *noh* and *nihon buyō*. Various insights from practitioners are given to explore how they perceive *kyojō* and the *iemoto* system.

The *iemoto* system

The *iemoto* system is structured in a strict hierarchy, and the *iemoto* holds absolute authority (Nishiyama 1982, Verley 1989). The term *iemoto* refers to 'the head of a family or head master transmitting the orthodox traditions of the school in the areas of learning, religion, or light cultural accomplishment and martial arts' (Kitano 1970: 5). In the *iemoto* system, students rarely have a direct lesson from *iemoto*. They have their regular lessons with local teachers who belong to the same *iemoto* system. Mori (1992) states that the *iemoto* system creates a hierarchy among students and teachers as shown in Chart 4.1:

There are five significant elements of the *iemoto* system: hereditary, absolute authority to *iemoto*, *kyojō* system, name taking system, and close master-pupil relationship (Nishiyama 1982. The details are follows:

DOI: 10.4324/9781003248682-5

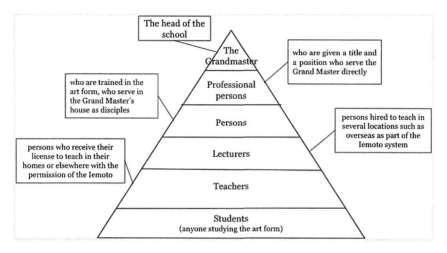

CHART 4.1 *Iemoto* system

Source: Mori 1992, Ōya 1993.

Hereditary system

Iemoto can be divided into *ie*, meaning house or family, and *moto* meaning origin or foundation. When translated into English, *iemoto* means grand master of the traditional art. This *iemoto* system, derived from the Japanese *ie* system, emphasizes the patrilineal line, the first son inheriting the family tradition and responsibilities. The position of grand master is also hereditary and generally passed down to the male heir. Thus, the first son receives training and the appropriate education from an early age (Nishiyama 1982. If there is no son, but there is a daughter, the *iemoto* family may organize an arranged marriage to let her husband carry on their family tradition. It is also not unusual to carry out an adoption if there is no heir; the Urasenke *chadō* grand master did not have any heir and arranged an adoption from the Matsudaira family who had substantial political and financial power during the Edo period (Sen 2000). There has also been an instance in which the first son declined his entitlement to being the heir. One of the tea schools' first sons was not interested in becoming the *iemoto*, and the *iemoto* decided to pass his position onto his second son. This hereditary system also applies to professionals who serve the *iemoto*. Kanze School from *noh*, recognized as the biggest *noh* school in Japan, has a Katayama family who have served the Kanze *iemoto* directly since 1704. Katayama Kurozaemon serves as the 10th generation of the Katayama family and supports Kanze style *noh* in Japan (Katayama n.d.).

While still prohibited in *noh* and *chadō*, *ikebana* has started to allow females to take the position of *iemoto* since the Second World War. The largest *ikebana* school, Ikenobo, has officially announced that the next *iemoto* is *iemoto's* daughter.

This hereditary system of *iemoto* restricted practitioners from being the head of a school. However, what if you did not agree with your master's method of teaching style? Within the field of *ikebana*, there is a certain level of flexibility in regard to teaching styles, as a result, around 300 *ikebana* schools have been established in Japan (Akitaken Kado Renmei 2007). Sixteen different *ikebana* schools are registered with the Akita *ikebana* association, four of which were established in Akita city (Terada 2021). One of the *chadō* practitioners, Chikako-san has highlighted the positive effects of a more relaxed *iemoto* system:

> When I heard that there were *iemoto* from Akita there, I was very surprised. You know *chadō*'s world. It is almost taboo to open their own school and claim ourselves as *IEMOTO*. It is amazing that *ikebana* is very flexible about this. I wonder if *chadō* should ever be flexible like *ikebana*. I don't think it will become flexible, do you? But I think this is a good idea to a certain extent. It means that if you are good at it, you can be the top of the art, right?

Although *ikebana* and *chadō* are perceived as the traditional arts, it appears that their approach to welcoming new schools seems to be vastly different.

Absolute authority

An *iemoto* holds the exclusive right to preserve a traditional form of art. Verley (1989:173) describes the privileges of this authority as follows: designating professional names to those attaining a certain level, admission to school and public performances, and granting permission to use certain tangible and intangible properties of the school. The *iemoto* is generally well respected, his voice is heard from the top to the bottom of the *iemoto* hierarchy. Furthermore, it is significant to recognize that the *iemoto*'s family members and financial supporters are also regarded as those who belong to the top tier of the *iemoto* system, and as such can also influence a school's decisions. It is worth investigating the extent to which an *iemoto*'s wife, daughter, or sister can influence his decisions, especially because of their status as female and the issues with perceptions of gender roles in Japanese society. Although the voices of the *iemoto*'s family members might be listened to by the *iemoto*, the grassroot level voices are rarely heard. It is uncommon for the *iemoto* to seek a beginner's opinions and advice, especially within the well-established schools (Kitano 1970, Nakagawa 1959, Ōya 1993).

A practitioner's behavior indicates how the absolute authority of the *iemoto* influences all interactions with him, and those in his presence. When Urasenke *chadō* school practitioners meet *iemoto*, they are trained not to look at him directly and instead tilt their head and body slightly forward while their hands are on the floor instead of on their lap, indicating that they are ready to serve him. One of the *chadō* practitioners' husbands observed the *chadō* Association and commented on the portrayal of *iemoto*:

FIGURE 4.1 Practitioners with *iemoto*
Source: Photograph by Mihoko Chiba.

> I think he is treated like a God. Whatever he says, they follow and agree. Listen to how they talk about him, Oiemoto-SAMA, Oiemoto-SAMA. I have known this *iemoto* through my wife for over 50 years, but I think they might be brainwashed! They donate a lot of money to him, which is my salary. They never criticize him.

Oiemoto-sama's 'O' and 'sama' are the polite honorifics used for a highly respected person. This comment emphasizes the amount of respect shown by practitioners for the *iemoto*, and they do not question his authority (Figure 4.1).

Kyojō *system*

The *iemoto* controls the teaching quality and its transmission by establishing a licensing system (*kyojō*) (Figure 4.2). The *kyojō* system is understood to require permission from the *iemoto* if one wants to receive the license to learn traditional arts, or a certificate pertaining to a certain level of practice. Nishiyama (1982) argues that this system contributed to the formalization of the traditional arts and organization of practitioners into a nationwide network. Practitioners generally do not take any official exam to move to the next level of practice. Instead, practitioners need a recommendation directly from their own teachers. If the teacher has observed a practitioner practice the specific procedures and

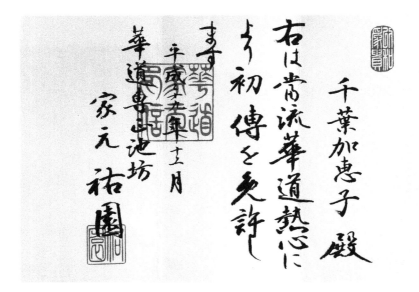

FIGURE 4.2 *Kyojō* picture

Source: Photograph by the author.

TABLE 4.1 *Kyojō* information

Name of School	Ikenobo	Sogetsu	Ohara	Ryuseiha
Kyojō Fee	No official information (Ikenobo n.d.)	No official information (Sogetsu n.d.)	No official information (Ohara n.d.)	10,000 yen – 60,000 yen (per one *Kyojō*) (Ryuseiha n.d.)

Source: Ikenobo, Sogetsu, Ohara, Ryuseiha n.d.

techniques for a sufficient amount of time, the teacher can advise the practitioner to apply for the certificate to qualify for the practice level.

Kyojō is generally recognized as a costly and timely procedure and is criticized when the length of time and cost of practicing are not accurately disclosed to the public (See Tables 4.1 and 4.2). The *kyojō* fee generally gets more expensive as it gets to higher levels. Furthermore, practitioners are required to pay annual fees to maintain their membership in the *iemoto* system, the high costs of materials needed for lessons, and a participation fee for concerts or exhibitions (Nomura 2008, Chiba 2010, Pecore 2000). Recently, some *ikebana* schools have made this information available to the public online. As one reaches a higher level of practice, the *kyojō* fee also increases; hence some practitioners purposely choose not to apply for recognition of reaching higher levels. One of the *ikebana* practitioners, Sato-*san* has commented about her choice:

TABLE 4.2 *Ikebana* three major schools

Name	Ikenobo	Sogetsu	Ohara
Year of Establishment	1462 (First ever appeared in the historical document, 'Hekizan Nichiroku')	1928 (After the first exhibition, it was broadcasted by NHK radio lectures)	At the end of 1800's (late 19th century)
Headquarters	Kyoto	Tokyo	Kobe
Generation Current *Iemoto*	45th generation, Ikenobo Sen-ei	4th *iemoto*, Akane Teshigahara	5th *iemoto*, Hiroki Ohara
Gender	Male	Female	Male
Kyojō System Character	Yes Ikenobo *iemoto* also works as a monk in Rokkakudo (a temple established by Shotoku taishi).	Yes The first *iemoto*, Sofu Teshigahara, believed that *ikebana* was a creative art. This school respects the 'originality' of the person.	Yes The first *iemoto* Ohara established the popular form of *ikebana* (Suiban and Kenzan). The current *Iemoto* works globally and socially such as using SNS. (Akitaken Kado Renmei 2007)

Source: Akitaken Kado Renmei 2007.

You see that I am nobody. I am just a retired person who wants to enjoy arranging flowers. I do not need any qualifications. Even if I was asked by my teacher about the *kyojō*, I say thank you, but NO thank you. I know it is expensive to get the *kyojō*. I live with my pension and I do not have lots of money.

Terada *Iemoto* from the *ikebana* School in Akita also commented that this approach to *kyojō* is quite common among practitioners who simply seek to learn *ikebana* as their hobby, but for nothing else such as a social position in the society.

Natori, *name-taking*

Once practitioners are members of the *iemoto* system, they are guided to assume *iemoto* as the father and other practitioners as brothers and sisters (Hsu 1975). Oya (1993) states that the *iemoto* system is based on a father-child relationship, and one of the main purposes of this system is to incorporate as many non-blood related members as possible. To emphasize this familial relationship, the *iemoto* system

FIGURE 4.3 *Ikebana's* name

Source: Photograph by the author.

also provides a naming system (Figure 4.3). Once practitioners reach to a certain level, they can receive a dedicated name. The name generally has one specific Chinese character from the school and another character from the practitioners' name. For specific occasions including *ikebana* exhibitions, or Japanese dance performances, these names have been used for the audience at exhibitions and dance shows (O'Neil 1984).

Master, senior disciple relationship

As described above, students have their regular lesson with their teachers in their local area. Class is generally scheduled for once a week for a couple of hours. There is a close bond that forms during the ongoing master-pupil relationship (Yamagishi 1959). Once practitioners decide to follow a specific master, practitioners are encouraged to be committed for a long time. While appropriate teachers are generally recommended through family members and friends, there is the notion that meeting a master is fate (*en*), and without questioning the relationship, practitioners start to practice. O'Neil (1984) states that these relationships are built on a fictive kinship framework and stress loyalty to one's master and to the grand master at the top. The relationship between practitioners is also close. Beginners tend to learn their basic lessons from their seniors, the elder sister disciple (*anedeshi*) or elder brother disciple (*anideshi*). For one of the tea classes, a teacher actually teaches once every three lessons, the rest of the classes are taught by *anedeshi*. *Ikebana* lessons are used as part of therapy in North America, and Backman (2013) states that patients not only receive healing through flowers, but also from the close relationship with their teachers.

So far, the characteristics of the *iemoto* system have been described, then what kind of different schools do we have in *ikebana*, *noh*, and Japanese Dance (*nihon buyō*)?

Different schools in *ikebana*, *noh*, and Japanese dance (*nihon buyō*)

There are three major schools in *ikebana*; Ikenobo, Sogetsu, and Ohara. The following table outlines the most important characteristics of each of the three major *ikebana* schools.

Noh

Noh is the style of play created by Kanami (1333–1384) and Zeami (1364–1443). It is a form of art with *mai* movement. *Mai* uses horizontal movements and uses sliding feet (*suriashi*). Performers regularly use *noh* masks, which are perceived as part of the performers' bodies (Konuki and Suzuki 1994, Mizutani 2014, Nogami 1929). Performers usually act as women, men, old men and ghosts with masks, and a simple stage is prepared with space for the musical accompanists to the side. This art form was protected by Toyotomi Hideyoshi, from the Tokugaku family, and is perceived as Japanese high culture (Yamanaka 2015). *Noh* practice and mask making were also made openly available for women after the Second World War (Endo 1999). There are five major schools, Kanze, Hōshō, Konparu, Kongoh, and Kita (Table 4.3).

Nihon Buyō

A form of art influenced by *noh* and *kabuki*, *nihon buyō* is *mai* and *odori*. *Mai* is more influenced by *noh*, emphasizing silence and staticness; *sei* movement uses the horizontal steps, *suriashi*, and feet slide as opposed to stepping. *Odori* is more active, *nihon buyo* combines these two elements. *Kabuki* was established for mass audiences during the Edo period (Brandon 1999, Bach 1995). However, women were prohibited from performing. Because some of the women still wanted to dance, *nihon buyo* was founded and allowed the participation of women (Hahn 2007). Regarding performance, *noh* and *kabuki* are still dominated by males, whereas *nihon buyo* is comparatively more inclusive of female performers. *Nihon buyō* has about 200 different schools in Japan. There are five main schools, Hanayagi (the biggest school), Fujima, Wakayagi, Nishikawa, and Bando School. The characteristics of each are provided in Table 4.4.

Noh practitioner Sakurako-*san*

Noh practitioners for over 10 years shared Sakurako-*san*'s perspectives on the *noh* and *iemoto* systems. Sakurako-*san* resides in the Kansai area and is in her 40s. She

TABLE 4.3 *Noh* schools

Name	*Kanze*	*Hōshō*	*Konparu*	*Kongoh*	*Kita*
Year of Establishment	Origin is Yūzaki-za of Sarugaku in 1337–1392 (Nanbokucho era)	Origin is Tobi-za of Sarugaku. 1464? (record from 1430–1451 was lost)	About 600, Yet no records until 53rd Iemoto.	Origin is Sakato-za of Sarugaku. Started from the Muromachi era	1586–1653 A son of an eye doctor, Shichidayu established
Headquarters	Tokyo	Tokyo	Tokyo	Kyoto	Tokyo
Generation Current *Iemoto*	26th *iemoto*, Kiyokazu Kanze	20th *iemoto*, Kazufusa Hōshō	81st *iemoto*, Norikazu Konparu	26th *iemoto*, Hisanori Kongō	Absent
Gender	Male	Male	Male	Male	–
Kyojō System Character	No official information Kanze is the biggest group of Noh (about 900 members).	No official information Original location was in Nara, as a relative of the Kanze family.	No official information First *iemoto* served for Shōtoku taishi. The oldest and most classic style.	No official information Only one family which has a headquarters in West Japan.	No official information First *iemoto* learned and leveled up various forms. Strongly supported by Hideyoshi and Ieyasu. (Yamanaka 2015, Ito 2021)

Source: Yamanaka 2015, Ito 2021.

TABLE 4.4 *Nihon Buyō* schools

Name	Hanayagi	Fujima	Wakayagi	Nishikawa	Bando
Year of Establishment	1848	About 1704	About 1893	About 1700	1775–1831
Headquarters	Tokyo	Tokyo	Kyoto	Tokyo	Tokyo
Generation Current *Iemoto*	5th *iemoto*, Jusuke Hanayagi	8th *iemoto*, Kanjūrō Fujima, *soke*, Kanemon	4th *iemoto*, Juen Wakayagi	10th *iemoto*, Senzou Nishikawa	11th *iemoto*, Minosuke Bando
Gender	Male	Male	Male	Male	Male
Kyojō System	Natori Exam –20,000yen for exam fee –10,000yen for venue fee	No official information	No official information	No public information (Only for students)	Light explanation. No cost information
Character	Biggest family of *nihon buyō*.	Dynamic movement, contrary to Hanayagi.	First *iemoto* learned in Hanayagi, but was excommunicated.	There are many branches from the original Nishikawa- ryu.	Uses 'Oyakata' or 'Danna' to call *iemoto*. (Hahn 2007)

loves Japanese traditional arts and practices, not only *noh*, but also *chadō*. One day, her friend, who practiced *noh*, invited her to practice with her and she started:

> My motivation was very simple. My friend brought a very cute advertisement to practise *noh*. I was living in Tokyo at that time, and I heard that my *noh* teacher was going to come from Kyoto and offer practicing *noh*. I thought this might be interesting, so I decided to go. I guess I still continue this *noh* practice, since I really like my teacher. He is a WONDERFUL teacher. And, *noh* is so attractive - the sounds, the movement, and the atmosphere! At first, you do not know what is going on, but once we practice, we will know the meaning behind it, and we will get attracted more and more.

Sakurako-*san* further shared that she practiced twice a month. Every lesson was a private lesson for about 30 minutes. The practice fee per month is 15,000 yen. Practitioners do not have to wear *kimono*, rather they sometimes wear trousers to practice *Noh* dance, *shimai*:

> Yeah, we can practice either singing, *utai*, or dance performance, *shimai*. I used to practice *shimai*, but now, I practice *utai*. In terms of gender roles in *noh*, I feel that some parts cannot be changed, for example, *utai*. I really feel that *utai* is for men. I simply cannot produce a deep voice like my teacher.

She stated that around half of practitioners are men and the majority of practitioners are senior. While the age background is similar to *chadō* practitioners, she commented that *chadō* and *noh kyojyō* style are slightly different:

> I felt that *noh kyojō* is only offered to those who really want to pursue it as a professional *noh* performer. I was never offered to apply for *kyojō* by my teacher. I guess he thinks that I do not want to be a professional performer.

From Sakusako-*san*'s interview, the characteristic of the *iemoto* system; the high cost of lesson fees was apparent. On the other hand, the *kyojō* style seems to be slightly different from other traditional arts including *ikebana*. *Ikebana* seems to open the *kyojō* to everyone who wants to pursue it further, and the *iemoto* rely on substantial income from it. However, *Noh iemoto* tends to earn their income mostly from their performance on the stage, rather than offering *kyojō* to practitioners.

Transparency and collaboration

Indeed, the *iemoto* system has been successful at sustaining traditional Japanese arts for centuries. The tight network in the *iemoto* system encourages better quality in the performing arts (Yamada 2017). By acquiring specific names from the *iemoto*, and fostering tight relationships, practitioners feel a sense of belonging. They can construct identities in a society in a positive way. However, as described above, there is some criticism toward the *iemoto* system. The Agency of Cultural Affairs (2018) has acknowledged the secretive nature of the *iemoto* system. Yamada (2017) argues that the *iemoto* system is an impediment to the development of performing arts in Japan: it ignores individual personalities, creativity, and differences.

In order to attract more of the younger generations and different groups, new, innovative ideas and initiatives could be considered and readied to be implemented. There are various ways to help cultivate a new generation of committed practitioners and increase general involvement with the traditional arts, including but not limited to: offering more detailed information regarding the licensing system to allow transparency and encourage inclusion; sharing knowledge and holding opportunities for collaborative work among different schools; and offering the positions of *iemoto* to women, a decision which aligns with the on-going gender equality social movement in Japan, and across the world.

Research questions

1. *Ikebana* costs money to be in a higher position and also just to learn it. Can it still be defined as a culture which is open to everyone?
2. What is the emotional relationship between the *iemoto* and his heir, and what are the heir's feelings towards his destined career?
3. Can we have creativity in the *iemoto* system?
4. The *kyojō* costs a lot, but may contribute to the preservation of *ikebana*. Could the licensing be an obstacle to having more practitioners, or is it necessary protection for its authority?
5. Is it possible to establish and continue running the *iemoto* system outside Japan?
6. Do you think it is acceptable to change the system of *iemoto*? If so, to what extent?
7. Will Japanese traditional arts die out without the *iemoto* system?
8. Can we apply the *iemoto* system to other things like management today?

Conclusion

In this chapter, we have examined the characteristics of the *iemoto* system. Information only available to practitioners gave a sense of exclusiveness. This style attracted some specific groups who preferred to have a different status from others. How to modify the *iemoto* system seems to be challenging, but should be discussed to preserve Japanese traditional art forms. This chapter also explored other traditional art schools in Japan and shared the voice of a female *noh* practitioner. In this secretive nature of the *iemoto* system, these voices are indeed crucial and offer a way to enhance their learning styles and systems among each traditional arts field.

References and further reading

Akitaken Kado Renmei. (2007) *Akitaken Ikebanashi* [Akita Ikebana History], Akita: Akita Kyodo Press.

Agency for Cultural Affairs. (2018) *Seikatsu Bunkato Jitai Haaku Chosa Jigyou Hokoku [The Report for Investigation in Life, Culture, etc]*. www.bunka.go.jp/tokei_haku sho_shuppan/tokeichosa/seikatsubunkato_jittai/pdf/r1403203_01.pdf (accessed 12 September 2021)

Backman, C. L. et al. (2013) 'Occupational engagement and meaning: The experience of ikebana practice,' *Journal of Occupational Science*, 20, 3. DOI:10.1080/14427591.2012.709954

Bach, F. (1995) 'Breaking the kabuki actors' barriers: 1868–1900,' *Asian Theatre Journal*, 12/2, 264–279.

Brandon, J. R. (1999) 'Kabuki and Shakespeare, balancing yin and yang' *TDR/The Drama Review*, 43(2), 15–53.

Chiba, K. (2010) *Japanese Women, Class and the Tea Ceremony: The Voices of Tea Practitioners in Northern Japan*, London: Routledge.

Cang, V. G. (2008) 'Preserving intangible heritage in Japan: the role of the iemoto system,' *International Journal of Intangible Heritage*, 3, 71–81.

Corbett, R. (2018) *Cultivating Femininity: Women and Tea Culture in Edō and Meiji Japan*, Honolulu: University of Hawai'i Press.

Endo, J. C. N. (1999) 'The contemporary female maskmaker: How females penetrated the male-dominated art of maskmaking.' *University of Hawaii Dissertations & Theses A&I.*

Hsu, F. (1963) *Clan, Caste and Club*, Princeton, New Jersey: Van Nostrand.

——— (1975) *Iemoto: The Heart of Japan*, New York: New York: Halstead.

Hahn, T. (2007) *Sensational Knowledge*, CT: Wesleyan University Press.

Ikenobo. (n.d.). *Ikenobo ni tsuite* [About Ikenobo]. www.ikenobo.jp/ (accessed 10 September 2021).

Katayama, K. (n.d.). *Katayama.* www.arc.ritsumei.ac.jp/k-kanze/ (25 November 2021).

Kitano, H. (1970) *Japanese American: The Evolution of a Subculture*, New Jersey: Prentice Hall.

Konuki, S., and Suzuki, M. (1994) 'Hyojyo ninchini oyobosu noh-men no kakudohenka no eikyo' [Impression change in the angle of view of 'Noh' mask on facial recognition], *Waseda University Human Science Research*, 7(1), 23–32.

Mizutani, S. (2014) 'Noh-men-no rekishi: Zoukeiteki shitenn-kara toraeta noh-men' [History of Noh masks: Noh masks from a modeling perspective], *Bulletin of Comprehensive Cultural Studies, Kyoritsu Women's University and Kyoritsu Women's Junior College*, 20, 39–49.

Mori, B. (1992) *Americans Studying the Traditional Japanese Art of the Tea Ceremony: The Internationalizing of a Traditional Art*, San Francisco: The Edwin Mellen Press Ltd.

Mori, K. (1971) 'A note on the modern iemoto system - reorganization of the iemoto system,' *Memoirs of the Graduate School of Meiji University Law*, 9, 222–234.

Moriya, T. (1980) 'The formation of the iemoto system,' *Bulletin of the National Museum of Ethnology*, 4(4), 709–737. DOI: 10.15021/00004540

Nakagawa, K. (1959) 'The readjustment of the iemoto system to the modernization of society,' *Soshioroji*, 6(4), 1–25. DOI: 10.14959/soshioroji.6.4_1

Nishiyama, M. (1982) *Iemoto Seidō: Nishiyama Matsunosuke Chosakun* [Iemoto system: Nishiyama Matsunosuke Selection], Tokyo: Yoshikawa Kobunkan.

Nogami, T. (1929) 'The non-expressionism of the noh mask,' *The Japanese Journal of Psychology*, 4(1), 1–15.

Nomura, N. (2008) 'The iemoto system and the development of contemporary quiltmaking in Japan,' *Textile Society of America Symposium Proceedings*, 119.

Ohararyu. (n.d.). *About Ohara School of Ikebana.* Retrieved September 10, 2021, from www.ohararyu.or.jp/about/

O'Neil, P. (1984) 'Organization and authority in the traditional arts,' *Modern Asian Studies*, 18(4), 631–645.

Ōya, Y. (1993) 'Structural features of iemoto system and "initiation of art" [Iemoto seidō ni okeru kouzouteki tokusei to ginou no denju],' *Nenpo shakaigaku ronshu* (6), 155–166. DOI: 10.5690/kantoh.

Pecore, J. T. (2000) 'Bridging contexts, transforming music: The case of elementary school teacher chihara yoshio,' *Ethnomusicology*, 44(1), 120–136.

Ryuseiha. (n.d.). *Hajimete no katahe* [For the first time visitors]. www.ryuseiha.net/information/ (accessed 10 September 2021)

Sellers-Young, B. (1992) 'Kanriye Fujima's adaptation of the iemoto system,' *Asian Theatre Journal*, 9(1), 71–84. DOI:10.2307/1124250

Sen, S. (2000) *Urasenke Chado Textbook*, Kyoto: Tankosha.

Sogetsu Foundation. (n.d.). *Sogetsu toha [About Sogetsu].* ww w.sogetsu.or.jp/ (accessed 10 September 2021).

Surak, K. (2011) 'From selling tea to selling Japaneseness: Symbolic power and the nationalization of cultural practices,' *European Journal of Sociology*, 52(2), 175–208.

Terada, K. (2021). Personal Communication 7 July.

Varley, P. (1989) 'Chanoyu: from the genroku epoch to modern times,' in Varley, P. and Kumakura, I. (eds.), *Tea in Japan: Essays on the History of Cha-no-yu.* Honolulu: University of Hawaii Press.

Yamada, K. (2017) 'Rethinking iemoto: Theorizing individual agency in the tsugaru shamisen oyama-ryu,' *Asian Music*, 48(1), 28–57, 151.

Yamagishi, T. (1959) 'Foundational discussion on iemoto system: One problem in sociology of art, [Iemoto seidō ni kansuru kisoteki kosatsu: geijutsu shakaigaku ni okeru hitotsu no mondai], *Tesugaku* (37), 27–56.

Yamanaka, R. (2015) 'Nohgaku no gendai to mirai [The present and future of Nogaku], *The Nogami Memorial Noh Theatre Research Institute of Hosei University*, 5, 5–16.

5

TEA FLOWER

The tea flower is the only element which is alive in the tea room arrangement. Sen no Rikyū (1522–1592) is recognized as the great master of *chadō*, and he arranged the flowers as if they were blooming in the field; the style is quite simple yet seems to have an in-depth meaning behind it. Sen (1974: 2) comments, '*Chabana* (tea flower) may not be a flower arrangement which has a *Zen*-like mission, but it certainly is a flower arrangement appreciated by a mind nourished with the Zen principles of simplicity and directness.' He further described the tea flower as the abstract expression of the self. In this chapter, readers will explore how the tea flower arrangements are different to the *ikebana* arrangements. In *chadō*, the flower arrangement is called *chabana*, and flower containers are called *hanaire*. This chapter further examines what kinds of flower containers are appropriate for specific occasions. It also explores the meaning behind *chabana* and flower containers for *chadō* practitioners, sharing insights from an interview with practitioners in Akita.

Chadō as composite art form

To examine *chabana*, it is important to understand *chadō* as the composite art form (*sōgōbunka*): it is composed of many elements including architectural design, the garden, utensils, tea and sweets, the meal (*kaiseki*), and flowers (Kato 2004, Chiba 2010) (Figure 5.1). Calligraphy scrolls and incense containers are arranged along with the flowers in the alcove. On the other side, the kettle, a fresh water container, and other utensils are placed in the typical tea room. All these elements have to be in harmony with each other, hence the flower arrangement is also humble compared with an *ikebana* arrangement.

DOI: 10.4324/9781003248682-6

FIGURE 5.1 *Chabana* arrangement with other utensils for New Year gathering

Source: Photograph by the author.

Sen no Rikyū (1522–1591)

As described briefly in the chapter on history, Sen no Rikyū was from the merchant class in Sakai, Osaka. He learned *chadō* from Takeno Jōo (1502–1555) and taught it to many, including Oda Nobunaga and Toyotomi Hideyoshi. Influenced by his master Takeno Jōo, he also appreciated *wabi sabi* aesthetics. Rikyū had a different appreciation for beauty to that of his student, Hideyoshi. According to Sen (1979), Rikyū heard that Hideyoshi was eager to see Rikyū's morning glories in his garden, and Rikyū invited Hideyoshi for a tea gathering. Hideyoshi was expecting to see numerous morning glories in the garden, but found that all of them had been intentionally cut away. Instead, Rikyū arranged just one morning glory in the tea room. Sen (1979) explains how this story about the morning glories describes how Rikyū perceived beauty in flowers: simple beauty. Rikyū also appreciated the *wabi sabi* aesthetics in *chadō*.

Sen no Rikyū stated the seven principles in *chadō*:

1. Make a delicious bowl of tea,
2. Lay the charcoal so that it heats the water,
3. Arrange the flowers as if they are in the field,
4. In summer suggest coolness, in winter warmth,

5. Do everything ahead of time,
6. Prepare for rain,
7. Give those with whom you find yourself every consideration.

Rikyū insisted that this fundamental part of *chadō* is the most important and difficult part to fulfill. As can be sensed from these examples, Rikyū appreciated the simplest of elements in *chadō*. One of the principles is about flower arrangements. Rikyū was the tea master who introduced an arrangement of flowers to the tea room (Anderson 1991). For Rikyū, the flower arrangement in the tea room should be as if it were blooming in the field.

Appropriate flower and container

Then, what kind of flower is appropriate for *chabana*. *Chabana* should be seasonal, flowers which can be seen in the field in that season or the month are appropriate for *chadō*. They should be unscented because incense is also arranged during the tea gathering, and the fragrances should not overlap (Mittwer 1974).

FIGURE 5.2 *Chabana* arrangement with a porcelain container

Source: Photograph by the author.

FIGURE 5.3 Bamboo container

Source: Photograph by the author.

Hanaire (flower containers) can vary in their formality, each pertaining to either *shin*, *gyō*, or *sō*, meaning formal, semi-formal, and informal respectively. Bronze and porcelain flower containers are recognized as the most formal (*shin*) utensils (Figure 5.2). Glazed pottery made in Japan, such as Seto, Tanba, and Karatsu, is defined as semi-formal (*gyō*). Bamboo, wicker baskets, and unglazed Japanese pottery including Bizen, Shigaraki lacquerware, and glass flower containers are considered as the informal (*sō*) flower container (Figure 5.3). It does not mean that the *shin* arrangement should be most respected, as even the most informal style expresses the refined flower arts (Sen 2000).

Tea flower as monthly

Seasonal changes are expressed by the flower arrangement in *chadō*. The appropriate flowers with a specially chosen *hanaire* are introduced in Table 5.1. The blooming season happens at different times depending on the region in Japan, but the figure below shows some typical seasonal flower arrangements.

TABLE 5.1 Tea flower seasonal arrangement

Seasons	Flower	Container
New Year	Willow	Bamboo
Winter	Green bamboo, acer palmatum (*tsukubane*), camilla, daffodil	Blue porcelain with crane neck style, style, bamboo, *ichijyugiri* style with turtle motif
Spring (Rikyū Memorial Service)	Rapeseed flower	Shigaraki ware
Early Summer	Iris	white and blue porcelain (*sometsuke*)
	Siebold's magnolia (*ooyama renge*)	Bizen ware
	Cypripedium japonicum (*kumagaiso*)	Metal ware (*karakane*) *shin* style
Summer	Japanese hibiscus (*mukuge*)	Cicada shape bamboo basket (*semikago*)
	Japanese morning glory	Bamboo basket
Autumn	Clematis (*tessen*)	Bamboo basket
Autumn	Bell Flower, bush clover and Japanese pampas grass	Bamboo basket favored by Sozen tea master
Early Winter (Robiraki)	Camillia, bamboo	Bamboo
	Winter sweet (*robai*)	Bamboo

Source: Sen 2000.

The camellias with leaves above their buds can be used for arrangements in the middle of winter; it shows as if the flower is protected by the leaves from the snow. In order to express the spring, camellias with buds sitting above the leaves are selected. This style is selected for the first tea gathering of the year, the first use of the fire pit (Mittwer 1974).

Appropriate place

Flowers can be placed on the *tatami* floor in the alcove, hanging on the wall, or hanging from the ceiling in the shape of a boat. For most occasions, they are placed on the *tatami* floor, and the board (*shikiita*) is generally placed underneath the flower container to protect the floor from damage. The flower arrangement in Figure 5.4 above is for autumn, and for this basket arrangement, the board is not used.

How to arrange *chabana*

First, select the appropriate seasonal flowers and containers suitable for the tea gathering. One, two, three, five, and seven flowers and branches can be placed as

FIGURE 5.4 Flowers with a basket for early autumn

Source: Photograph by the author.

odd numbers are preferred. For *ikebana* arrangements, it is encouraged to arrange each flower and branch, from the main branch, second branch, and third branch. In *chabana*, practitioners choose the flowers that they want by arranging them one by one in their hand first, and finally, placing them in one motion into the container (Mittwer 1974).

The meaning of tea flowers to practitioners

Flower shops in Japan generally do not offer flowers which bloom in the field, so how do practitioners prepare flowers? Some practitioners order from special flower or garden shops. The Urasenke tea school has their own nursery in the suburb of Kyoto and raises their own tea flowers. Other practitioners devote themselves to gathering field flowers from other practitioners, as well as raising flowers of their own. Kadowaki-*sensei* living in Akita city talked about practitioners who have their own gardens and tend to plant their flowers there, and shared how she gathered her own tea flowers. In her 70s, Kadowaki-*sensei* has been engaged with *chadō* for more than 50 years, and invited me to her garden. It was at the beginning of May, and the garden was full of new buds from the tree branches to the ground. She slowly walked through the garden and described how she got her tea flowers one by one. Some of the tea flowers were protected from other plants on the ground by wires. Kadowaki-*sensei* commented:

> You see, tea flowers are the flower blooms in the field. They look like weeds for some people who are ignorant like my husband and my father. They think this precious tea flower is a weed and get rid of it! So, I protect it from these ignorant people and make these protections. They really do not know how much time and effort has been put into these flowers, not only mine and my mothers, but also from the person who fetched this flower for me.

Kadowaki-*sensei* further commented that some camellias are easy to be found in the local nursery, however, the rest of them are shared by other practitioners. Yamada-*sensei* describes:

> When we moved to this house, we did not have any *chabana*, so my mother and I asked other practitioners, locals who live or visit mountains to bring some tea flowers from their gardens, fields, and elsewhere, and every time when they brought them, we checked if these flowers could be shared. For example, some flowers, especially flower trees can be shared by growing roots again. We placed the branches in the pot with soil, and sometimes it failed, but sometimes it worked!

She further describes:

> It is also not easy to raise tea flowers in the garden since we have to create a similar condition to the field. For example, the amount of sun and fertilizer. To make the soil condition similar to the field, we keep the fallen leaves in bags for four years and return them back to the soil. We also add ashes from the charcoal to make the soil better. One of my students even goes to the mountain every year and fetches fallen leaves as fertilizer. We make sure that tea flowers do not get too direct sun by planting big trees to provide enough shade for tea flowers just as if they were in the field.

She took me to her tea garden and explained how each plant came to her place.

Miyakowasure:

> So, this one came from one of my practitioners, Sawaki-*san*. We placed it in the water for a while and the root came. Later on, we placed it in the pot. It will bloom soon with light purple. The flower itself is so small,

FIGURE 5.5 *Ooyama Renge* with formal (*shin*) style

Source: Photograph by the author.

but the purple color gives us a statement. The name of this flower is interesting, right?

Miyako means the city capital and *wasure* means forgetting. It is said that this flower was named after one of the emperors who was exiled to Sado during the Kamakura Period (Figure 5.6). He saw the color of this purple flower which reminded him of his nobility in his capital city. He decided to forget his city by using this flower for his poem (Mittwer 1974).

Kumagaiso:

One of my carpenters knew that I do tea and fetched it from the southern part of Akita. He collects wild flowers from the mountain. He also gave me good guidance on where to plant it, like I should plant it underneath the tree in a rather dry place. I was so grateful for his advice. We were able to have flowers in the following year.

FIGURE 5.6 *Miyakowasure* with a basket

Source: Photograph by the author.

Kumagaiso is the wild orchid which has light purple blooms. The name derived from the *horo* (floating cloth that *samurai* warriors used to wear on the back as protection from arrows) that Kumagai warriors used to wear (Mittwer 1974). The name of the tea flower itself sounds very poetic.

Shirane Aoi:

Oh gosh, this is such a precious one! Ozawa-*sensei* brought it from Ani, deep in the mountains. He bought it from the wildflower shop in Ani. I really like this subtle color and since this flower blooms from the vine, we use this flower for arrangements from the wall. I have the white one and purple one, and both are very pretty.

Taitsuriso:

Ozawa-*sensei* brought it for me. We had it before, but it was gone. You see, wildflowers easily disappear. I heard that our garden is not like the environment in the mountains. I think this heart shaped flower might become popular among young girls since it is very pretty.

Kuro Yuri:

My mother and I went to the mountain and fetched it. We heard that if we went there, we could get it. We rarely went to the mountain. We really did our best! On that day, we had a snake close to that flower. We screamed and shouted, but this one is such a sensitive one, it stopped blooming. I hope she will come back next year.

This flower is a relatively small black lily flower that grows in the northern part of Japan. Kadowaki-*sensei* also shared the belief among the Ainu that if this flower is placed next to the person's pillow without being noticed, the love they hope for comes true.

Odoriko-so:

We got it from the rice field. Nowadays, we tend to use chemicals, so it is rare to get this flower from the rice field. We may only be able to get this through shops. I love the sound of this flower's name. The flower shape is similar to the dancer who wears a straw hat, which is why it is called *Odoriko-so.*

Kadowaki-*sensei* further explained how practitioners make an extra effort to collect precious tea flowers for the tea gatherings or tea seminars. She also went deep into the mountains to fetch a specific flower. After collecting flowers, water was constantly sprayed onto them to keep them in good condition. She commented about the reaction from other practitioners:

It was nice that other practitioners appreciated flowers at a tea gathering. Most of them know how hard it is to acquire suitable flowers for a gathering and keep them in the best condition, not fully bloomed. *Chabana* is alive and so sensitive, so to present this flower for her best moment needs considerable effort. I get so happy and emotional when other *sensei* comment on our flowers.

From Kadowaki-*sensei*'s interview, we can see how she devotes herself to tea flowers. She treats flowers as her precious gifts. Practitioners seem to spend considerable time raising and maintaining flowers for tea gatherings, with the time and effort put into making the soil for the specific flowers, it seems that this process is also considered spiritual training (*seishin shugyo*).

When I had an interview with Kadowaki-*sensei*, I also had an opportunity to interview another practitioner, Matsuhashi-*san* who knows well about

Kadowaki-*sensei*'s mother. Matsuhashi-*san* shared her memory about Kadowaki-*sensei*'s mother.

> She passed away 5 years ago, but I remember so well. She loved *chabana*. Whenever she had spare time, she was always in the garden fixing the soil for *chabana*. Her hands were always dirty! She first placed all the *chabana* in the pot to make sure it was warm enough and healthy. She took pictures as they grew. Once she was sure that they were strong enough to be outside, they were located at the right spot in the garden, they were well looked after by *sensei*…. Now look at them, they are all well protected by these big trees. There is enough shade for them to grow, enough sun to grow.

Indeed, Yamada-*sensei*'s tea flowers were well protected by trees as if we were in the field. Practitioners' devotion to tea flowers was felt not only through the comments from the interview, but through the aged trees and moss from her garden.

Meaning of flower containers

So far, I have described how practitioners pay great attention to flowers themselves. However, they also pay considerable attention to selecting flower containers. Kadowaki-*sensei* explained:

> I personally love this container, *zorori*, but it does not mean that I can use this all the time. I have to select the flower container to match the flowers and seasons. Not only that, but tea gathering always has a theme, so we try to match the container along with the theme, too. I think this is why we have so many different styles of containers in tea. Not only that, but we have to be aware that we have to select the appropriate utensils for appropriate tea gatherings. This means that sometimes, it is polite to select relatively well known artist's utensils for specific tea gatherings.

Kadowaki-*sensei* continued:

> So, for our local tea gathering, we selected a shigaraki ware flower container made by Sadamitsu Sugimoto - of course the certified one. I adore this container, the mild pink and creamy color and the unbalanced shape. Its rough surface conveys a sense of warmth. Since it looks relatively heavy (*doshiri*), it goes so well with the branch arrangement. On another occasion, we selected a Rikyū favored style bamboo flower container. This one goes well with every arrangement, but since it is simple, a single camylia arrangement is just beautiful, as this arrangement expresses the simple beauty.

FIGURE 5.7 Certification boxes

Source: Photograph by the author.

Kadowaki-*sensei*'s stories tell that practitioners also put great effort into selecting flower containers by considering their appropriate status: for well-recognized tea gatherings, practitioners tend to select a flower container which has the *iemoto*'s signature on the box. With the signature on the box, the price increases (Figure 5.7). Kadowaki-*sensei* commented that she paid 400,000 yen for her shigaraki flower container, and 800,000 yen for the bamboo container. These prices appear to contradict the *wabi sabi* philosophy, however, this is another phenomenon in traditional art culture which seems to be related to status and social class distinction in Japan (Chiba 2010).

Research questions:

1. Why did Sen no Rikyū focus on the beauty of flowers in nature?
2. In what ways do the tea flowers mentally play a role in a tea gathering?
3. At what point do people think tea flowers are beautiful?
4. Why is an odd number preferred when you place branches or flowers?

5. What are the most common flowers used in *chabana* and is there any connection with the meaning of flowers?
6. If you had a tea gathering, which flowers and *hanaire* would you select?
7. As one of the elements of *chadō* includes using flowers, are there any other Japanese traditional arts that could incorporate flower culture into the art form?
8. In what way(s) is *chadō* akin to the phrase, 'Stop and smell the roses?'

Conclusion

In this chapter, we explored the characteristics of flower arrangements for *chadō*. The flower arrangements are simpler than the *ikebana* style, including the amount of flowers, flower colors, and containers. Tea flowers are even simple in terms of their scents. They hardly have any fragrance in order to avoid any conflict with the incense in the tea room. Tea flowers may be simple, yet practitioners make timeless devotion for these flowers. As described above, it takes years to grow tea flowers in the most desirable way. It appears that tea flowers mean their way of training – the way of life to *chadō* practitioners. On the other hand, tea flowers play a significant role in expressing the *wabi sabi* aesthetics in tea gatherings: appreciating the moment in a flower's life, recognizing that this is not eternal.

The information mentioned above is a desirable arrangement. Then, what happens if we are outside of Japan and cannot access the flowers and containers listed above? Flowers which bloom in the local fields overseas can be easily used. Simple flower containers or other similar containers as alternatives could also be arranged. Local artisans' work can be used for flower containers, or we can easily make our own containers by cutting bamboo like Rikyū. A flower container does not have to be expensive, nor well-known; it can be simple and humble just like tea flowers.

References and further reading

Anderson, J. L. (1991) *An Introduction to Japanese Tea Ritual*, New York: State University of New York Press.

Chiba, K. (2010) *Japanese Women, Class and the Tea Ceremony: The Voices of Tea Practitioners in Northern Japan*, London: Routledge.

Hamabata, M. (1990) *Crested Kimono: Power and Love in the Japanese Business Family*, Ithaca: Cornell University Press.

Hsu, F. (1963) *Clan, Caste and Club*, Princeton, New Jersey: Van Nostrand.

Hsu, F. (1975) *Iemoto: The Heart of Japan*, New York: New York: Halstead.

Kato, E. (2004) *The Tea Ceremony and Women's Empowerment in Modern Japan*, London: Routledge.

Mittwer, H. (1974) *The art of Chabana: Flowers for the tea ceremony*, Tuttle Publishing.

Sen, S. (1979) *Tea Life, Tea Mind,* [Foreign Affairs Section, Urasenke Foundation, trans]. New York: Weatherhill.

——— (2000) *Urasenke Chado Textbook*, Kyoto: Tankosha.

6

JAPANESE GARDEN

There are different styles of Japanese garden (*nihon teien*) in Japan. This chapter explores how flowers have been utilized in Japanese gardens. It also examines the in-depth meaning of each garden and how numerous types of techniques are established to maintain the beauty in Japanese gardens. This chapter will also share interviews from a highly skilled gardener (*niwashi*) and his family members to examine their training style, work ethic, and their thoughts on the future for Japanese gardens. Japanese talk about gardens as *niwa*, or *teien* in Japanese. *Niwa* connotes a garden for a private household. Whereas the *teien* tends to be used for the well maintained formalized gardens, most of which are open to the public. Japanese *teien* and the background meanings are the focus points in this chapter.

Character of Japanese garden

Around 80 percent of our land in Japan is covered with mountains, only around 20 percent of the land is available for habitation. This means the land that Japanese can afford to have as gardens is relatively limited. Japanese gardens may not be large, but there are numerous techniques and skills being used to make us feel comfortable (Zhou 2013). Most of the Japanese gardens are composed of stones, water, and evergreen trees. Flowers can be used, however, the garden usually maintains an image of green rather than multiple colors from flowers. It appears that most of the Japanese gardens try to represent all nature in Japan, and the Japanese perceive this whole nature as generally green rather than colorful flowers (Habib *et al* 2013, Naka 2009).

The concept of a garden came from China and Korea. In the Nara period (710–794), there was already a record of a garden used for festivals and parties as well as ceremonies. As with *ikebana*, Japanese gardens are influenced by Buddhism and Shintoism. Amenism related to Shintoism believes that the gods

DOI: 10.4324/9781003248682-7

exist in water and stones, so these components are often used. Japanese gardens comprise numerous symbols, and these are not only the symbols of gods (Stauskis 2011). Streams of water symbolize a river. The still water sometimes represents a lake or a pond in the local area. A rock in a pond is interpreted as an island, a rock by the edge of the water can be perceived as the rough coast, and numerous rocks arranged from top to bottom may represent the waterfall (Naka 2009). The following are some different styles of gardens.

Pure land garden

Pure land gardens were established among aristocrats in the Heian period (794–1185). This garden is influenced by Buddhism especially from the Jodo school, and it expresses the idea of pure land in the afterlife. Ponds, islands, bridges, waterfalls, and Phoenix Hall are used to express the image of pure land in the concept of the Jodo school. Byodoin Temple in Uji, Kyoto is well known for this garden style and was also registered as a World Heritage Site in 1994 (Naka 2009) (Figure 6.1).

Shoin *style garden*

Along with the *shoin* architecture, which introduced the concept of alcove, the garden attached to this architecture was introduced during the Kamakura period

FIGURE 6.1 Byodoin Temple

Source: © Byodoin.

(1192–1333). The size of the garden is much smaller than the pure land garden. The Jisho-in in Kyoto is well known for this *shoin* style garden. The garden was designed to appreciate viewing from indoors (Naka 2009). Daigoji Sanboin Garden in Kyoto is also recognized for this style, and it was registered as a world heritage site in 1994.

Dry or rock (karesansui) *style garden*

During the Muromachi period (1185–1333), the concept of rock gardens emerged. This garden style uses rocks and stones instead of water or greenery to express the beauty of mountains and water flow. Naka (2009) argues that this garden style emerged from the financial impediment following the Battle of Ounin (1467–1477). This garden style aims to provide a moment to practice meditation, and it represents the concept of emptiness. Monks are encouraged to meditate while raking the stone. Visitors are also encouraged to calm themselves and practice meditation. Ryoan temple in Kyoto has an interesting rock garden (Figure 6.2). There are 15 stones arranged in the rock garden, however, it is intentionally arranged so that the visitors can only see 14 rocks, making them seek the imperfection of beauty (Naka 2009).

FIGURE 6.2 Ryoanji

Source: Photograph by Bhikhu Sagarananda Tien.

Tea garden

Tea garden is called *roji*. This place offers guests a space to detach from their busy daily lives, providing a moment to be calm in their minds and prepare themselves to have tea in the tea room. The typical Japanese tea garden is generally green with fewer flowers. The size is compact, as the main purpose of this dedicated space is to help make guests feel calm as they walk along the stepping stones before tea (Figure 6.3). Guests enter the gate of the outer *roji*, and they wait for other guests at the waiting area. After being guided by a host/hostess, the guest goes to the water basin (*tsukubai*) to purify themselves before entering the tea room (Sen 2000).

Sen (2000) comments that tea gardens started at the time of the Muromachi period simply with several rocks and shrubbery to begin with; Rikyū tended to have flexible ideas towards the design of a tea garden, but emphasized the *wabi sabi* aesthetics. Later on in the period, Furuta Oribe (1544–1615) and Koburi Enshū (1579–1647) enhanced the concept of a boundary between the inner and outer garden (Sen 2000).

FIGURE 6.3 Tea garden

Source: Photograph by the author.

Stroll style (kaiyushiki) garden

This style was created during the Edo period (1603–1868) in order to enjoy walking around the garden. It was favored by the *samurai* class in Japan to represent their status. Visitors appreciate the garden view from the paths and bridges. Some of them have tea houses and resting places. To make the garden feel as if it is bigger or to appreciate the beautiful view from the scenery, some gardens utilize a technique called *shakei* and borrow the view of the local mountains in the background (Zhou 2013). Naka (1990) comments that there were numerous feudal loads' gardens in Tokyo during the Edo period, and their gardens also had another purpose – to store large amounts of water in case of fire.

Jyoshitei, located in Akita city, known as the garden for Satake lords, contains elements of a stroll garden. It was built between 1688–1704 by Satake Yoshimasa (Figure 6.4). A small-sized waterfall and stream represent the wishes from Yoshimasa that one should be a continuous learner for the rest of one's life as Taoism philosopher Koshi emphasized. Along with the stream, visitors appreciate the 15 different features including unusual rocks, lanterns, and a tea house. This garden is also designed as a viewing garden to be appreciated from inside of the resting house on the property. Jyoshitei renovated it and opened it up to the public in 2014. The rest house in the garden is regularly used for cultural activities including *Ikebana* exhibitions (Akita City Board of Education 2010).

Former Ikeda family garden

The Ikeda family was recognized as one of the three most powerful landowner families in the Tohoku area before the land reform. The garden was designed by the first landscaper in Japan, Nagaoka Yasuhei (1982–1925). The garden is a stroll style one around a large pond, and it covers an area of 42,000 square meters (Figure 6.5). Rocks, a waterfall, and a stone lantern measuring four meters in diameter are some of the highlights in this garden, along with the beautiful backdrop of local mountains and rice fields. This garden was registered as a National Place of Scenic Beauty in 2004 (Daisen City Board of Education 2010).

Tsuboniwa *style garden*

It is a significantly small sized garden started in the merchant area of Kyoto. This *tsuboniwa* has been useful to welcome light and breezes to a house where there is insufficient space between neighboring houses. Tea gardens influenced this garden style, as most of them have the elements of stone lanterns and stepping stones. Naka (2009) argues that this style can be developed in the future to adapt to urbanization.

FIGURE 6.4 Jyoshitei

Source: Photograph by the author.

Western style garden

After the Meiji period (1868–1912), the Western style garden influenced Japanese gardens. Furukawa Garden in Tokyo is well known for this style. The garden is arranged with a blend of Japanese and Western aesthetics: Some parts of the

FIGURE 6.5 Ikeda family garden. Copyright Daisen City

garden are well designed with symmetric flower beds, and other sections of flowers are arranged asymmetrically to balance with the Japanese style garden (Naka 2009).

Entsuin temple garden, based in Matsushima, Miyagi prefecture, represents the beauty of Japanese gardens and western flowers; roses. The lord of the Miyagi area, the Date Masamune lord, dispatched an envoy to Europe at the beginning of the Edo period (1603–1868) and brought back the rose seeds. When the envoy returned to Japan, the government changed its policy and introduced restricted trade (*sakoku*). The head of the envoy, the Hasekura family, was forced to disband due to the fear of Christian influence, but the roses grew successfully and have continued to bloom in the temple garden since then. It is substantially rare to see rose flower arrangements in the Japanese graveyard garden, and it is recognized as the oldest rose garden in Japan (Miyagi Prefecture Board of Education 2021).

There are some other gardens that may not be categorized as above: moss garden. A well known moss garden is located in Kyoto, Saiho-ji. Moss may not be appreciated in other societies, however, as described in the Japanese national anthem, it is perceived as the symbol of the *wabi sabi* concept in Japanese gardens in relation to an appreciation of time and simplicity (Naka 2009). The owner of the moss garden in Saiho-ji encourages visitors to have spiritual discipline before visiting the garden by booking in advance with only a postcard, visiting the site by public transportation or foot only, and copying a transcript before entering the garden.

TABLE 6.1 Three major gardens

Name	Kairakuen	Korakuen	Kenrokuen
Year	About 1842	About 1687	About 1676
Place	Ibaraki Prefecture	Okayama Prefecture	Ishikawa Prefecture
Founder	9th Mito feudal lord, Tokugawa Nariaki (a biological father of the 15th and the last Tokugawa Shōgun Yoshinobu)	2nd Okayama feudal lord, Ikeda Tsunamitsu	5th Kaga feudal lord, Maeda Tsunanori
Features	Shortly shifted ownership to Ibaraki prefecture as Tokugawa park for the public.	Noh stage, rice field, tea plantation	Philosophy of Shin-sen thought Tea house

Source: Naka 2009.

Three major gardens

There are three well known gardens in Japan, Kairakuen, Korakuen, and Kenrokuen (Table 6.1). It is even a common belief that we should not die before visiting these gardens. Korakuen in Okayama has a tea plantation where guests can smell the beautiful aroma of roasting tea leaves and purchase tea in the garden (Collinge 2021). Kenrokuen based in Kanazawa is well known for its wintery view (Figure 6.6). To protect trees from snow by not bending their branches, there is a specific protection technique, *yukizuri*. It is made from straw string and visitors also admire the beauty of *yukizuri* with its snowy view. Kenrokuen was first opened to the public in 1874. Ever since, it has attracted numerous visitors from different backgrounds. Eighty-eight percent of tourists in Kanazawa city, including the younger generation visit Kenrokuen (Kanazawa Tourism Bureau 2020).

Japanese gardens outside of Japan

The Japanese garden provides a tranquil moment. It is the way to see *wabi sabi*. Suzuki, Kuroda and Aso (1995) commented that Japanese gardens are more popular outside of Japan. According to Suzuki (2007), there are around 350 Japanese gardens outside of Japan. Japanese gardens are sometimes created or donated to express the friendship between partnership cities or countries. The Japanese Tea Garden in San Francisco, the Portland Japanese Garden in Oregon, the Nitobe Memorial Garden in Vancouver, Canada are such examples.

FIGURE 6.6 Kenrokuen

Source: Photograph by the author.

Differences from the western garden

Japanese gardens follow a different aesthetic to western gardens. As described above, Japanese gardens are influenced by Buddhism and Shintoism. Shintoism believes that there are deities in nature and its elements, and Buddhism believes that we as humans are part of nature, so Japanese gardens try to represent nature itself (Isoya 1984, Naka 2009). On the other hand, western gardens have historically been influenced by the Christian idea that all things on the earth are God's creations and human beings are the only special creation among them, thus, man alone is a special creature that is superior to nature (Ito 2002, Watanabe 2005). Consequently, western garden creates a landscape which expresses order and formal beauty. For instance, the Palace of Versailles Garden in France expresses symmetric and geometric beauty. The hedge is trimmed with a straight, round, or curved line to express the beautiful artificial line. Numerous colors are arranged in flower beds to express that the garden is looked after well. The water is often organized from the bottom to the top as a fountain to represent the beauty and strength of manpower. Statues are used to enhance western garden beauty.

Japanese gardens utilize water and other elements differently from the western gardens. Water is generally used as the representation of nature, and so follows the natural flow of top to bottom in Japanese gardens. Carp (*koi*) fish bring another charming dimension of naturalness to Japanese gardens. The Japanese gardener trims the trees with specific techniques to ensure the trees still look as natural as possible (Naka 2009). Much like *ikebana* arrangements, most Japanese gardens are also asymmetrical. *Niwashi* are trained in many techniques to carefully and skillfully express the beauty of the garden; they not only maintain the garden but they also design the garden. The following section explores the narratives from the *niwashi* in Japan.

Yohei-*san*

Yohei-*san* is a nephew of a *niwashi* and has been working for his uncle for about 10 years. He commented that at first, he did not want to be the *niwashi*. He said that it had an image of not being cool. However, as he started to help his uncle when he entered high school, he started to get interested in his work. Yohei-*san* left his town for his training under the other *niwashi* and came back to work for his uncle. He explained about his training time:

> Well, it is quite interesting. Since it is not a school, we do not get detailed teaching from our boss (*oyakata*). We are simply encouraged to observe what he does. There is no atmosphere in which to ask questions. If we did, I think he would think we were cheeky (*namaiki*). So, first we simply observe and copy him, and sometimes, he explains to me why we do a process. I guess I was fortunate since he knew that I was planning to work under my uncle in the future. He gave me generous guidance. I certainly owe him, and I would like to repay him with my work.

After his long training, Yohei-*san* started to be offered the position of designing Japanese gardens. He first talked about the stepping stones that he designed.

> This looks all natural, but we make considerable effort to make it natural. First, we choose stepping stones. There are so many different stones from the rivers or from the mountains. All of them have different shapes and roughnesses, and we try to select the appropriate ones for every step. At first it all looked the same, but it looks all different to me now. It is thought that when we place stone with a horizontal line, it is beautiful, but if we follow this aesthetic, it becomes very uncomfortable to walk. So, we need a balance, and we walk on them so many times to try to check their practicality and aesthetics. Then we have to select trees and plants. The stones do not grow and stay as is, but other plants and trees do. So, when we purchase the plants, we have to imagine the future garden and how it grows in combinations with spaces. We do not want to have a bushy garden, we have to provide enough space for lights and other places to breathe.

Yohei-*san* also admired the stone lantern that he ordered from Yame-city in Kyushu. He pointed out that the stone from Yame absorbed moisture, so moss could easily grow on it and add the *wabi sabi* taste to the garden. He continued about other techniques:

> I always want to improve my skill. We make everything from scratch – these bamboo fences, too. Apparently, we can see the *niwashi's* quality by looking at these fences. I do my best to provide the best quality work to my client. I like my work and am proud of my work.

Yohei-*san* seemed to be proud of his work and concluded that he needed to pass these skills on to younger generations. Similar to Yohei-*san*, Masao-*san* also appears to start learning his in-depth technique through numerous observations.

Masao-*san*

Masao-*san* has been a *niwashi* for about 30 years. For most of his life, he has spent his time outside, so he is quite tanned. He is originally from Akita city, but he was working as a *niwashi* in Tokyo for decades and looked after numerous Japanese gardens there. However, he had a serious accident. He fell from a tall tree while he was trimming it, and had to have brain surgery. His mother thought he would never survive and recover. Fortunately, he recovered after a long rehabilitation in Akita city. First, everyone, perhaps including him, thought he would never go back to the same work. He took other occupational qualification classes and exams for other options. However, he is working as *niwashi* again in Akita city. His mother commented, 'I felt that *niwashi* is his life. He loves gardening. It is not well paid work, BUT he loves it. He is proud to be a *niwashi*.' Indeed, when I first met him, I could see how he loved gardens. He quietly stood in the garden and looked at the rocks, leaves, and grounds for a long time (Figure 6.7). Masao-*san* could immediately identify the rocks in the garden, and he commented with an excited voice:

> You see these rocks are from Kishu. These are blue stones from the Kishu area. This white pattern shows that it is aging. How precious! You see the stone taste changes with age, and it expresses different beauty. This garden gives great emphasis to the five element stones which are low vertical stone *(reishōseki)*, statue stone *(taidōseki)*, flat stone *(shintaiseki)*, arching stone *(shigyōseki)*, and recumbent ox stone *(kikyakuseki)*. These stones go very well with the blue maple trees *(aokaede)* and camila *(himeshara)* related trees.

I asked him why he wanted to be a gardener. He said:

> When I was working at the construction site, I saw my future gardening colleague working next to us. Their way of working was quite elegant, and

FIGURE 6.7 Rocks

Source: Photograph by the author.

I thought I would like to work with them. So, I applied for the position, and this is how I started. When I started, I did not know anything. My boss took me everywhere, and I learned by observing them. He rarely taught me what to do. I guess this is what we say, 'steal the skill (*waza wo nusume*).' I think he, my boss, wanted me to observe and learn his skill.

Masao-*san*'s mother also shared similar content in relation to his training style in Tokyo.

When he was in Tokyo, his master was very nice. He trained him well. He took him to all different styles of gardens and simply let him watch what he did and let him copy. I guess that was the style to express his care to him.

Masao-*san* continued:

But you see, once you learn all the skills, you feel kind of a dilemma with your boss. He wants you to pursue his own way, but I sometimes think there are some better ways. For example, for me, the trimming style should be flexible considering the residents' lifestyle.

Masao-*san* leaves home at 6AM and comes back home around 7PM, which is a long time. He comments that he is the youngest *niwashi* in his company, so he covers the places where the others cannot. He comments that there are not many young people working for this business. He further commented on gardening business in Japan as follows:

So, you see, nowadays, most of our work is not about Japanese gardens. It is rare to see companies only focusing on Japanese gardens. Why? Because there are hardly any people who want to make new Japanese gardens in Japan! I have not heard of any for a while, and they are seeking status by placing five huge rocks. Having said that, these Japanese garden cultures will not disappear. These gardens will last 300 and 400 years. I feel that all these famous Japanese Gardens, like Katsura Rikyū, Adachi Museum Garden will last forever, and our skills will be passed down to the next generation.

It was common to have *sumikomi* style where the apprentices live with the master until the person learns the skills. The apprentices were not paid, but were offered meals and a place to stay, generally in the same house with the master. While this *sumikomi* style is no longer common among *niwashi*, it appears that their learning style has not changed. Their learning style is similar to traditional arts training, observation, imitation and repetition. He further feels that there is less emphasis on Japanese gardens in Japanese society. However, it was interesting to see that he seems to feel less pressure that Japanese garden skills might die out compared with other traditional arts including *chadō* and *ikebana*.

Research questions

1. By rearranging natural materials in a desired manner, building a garden is, in some sense, like *ikebana*. Do you think we can still appreciate Japanese gardens as something natural?
2. Should we learn the history and the types of Japanese gardens rather than simply visiting them as site-seeing spots?

3. Do gardens still have the same functions (socialization, recreation, relaxation, and firefighting, etc.) today?
4. Why do Japanese gardens try to represent something with other materials such as representing water with stones?
5. Why does the Japanese style have few flowers?
6. What roles are Japanese gardens playing for the Japanese people today?
7. The concept of pocket gardens (*tsubo-niwa*) isn't new, but the way they have developed in the modern era is (e.g. *tsubo-niwa* in balconies). How do you think Japanese gardens will develop in the future to adapt to urbanization?

Conclusion

This chapter explored the meaning of Japanese gardens and how numerous types of techniques are established to maintain the beauty in Japanese gardens. We learned from the interviews that these *niwashi* acquire their precious skills from similar training styles with other artisans and practitioners in Japan. Most gardens in Japan were created to express status in society. This tendency was significantly apparent before the land-reform in 1946. Some of the ex-land owners or other middle class owners may have created Japanese gardens until the 1980s, however, as mentioned before, it is rare to encounter Japanese families arranging Japanese gardens at this time. Most Japanese housing styles are becoming more western, and so is their garden style, including the *tsuboniwa* style. Contrarily, Japanese gardens, even miniature rock garden kits, have been popular in western societies. We encounter numerous examples of cultural knowledge from different countries, and might think other cultures are more attractive than our own; the grass is always greener on the other side of the fence.

References and further reading

———— (2000) *Urasenke Chado Textbook*, Kyoto: Tankosha.

———— (2012) *Japanese Garden Design*, North Clarendon, VT: Tuttle Publishing.

Aida, A., Chang, K., Suzuki, M., and Taniguchi, S. (2003) 'Research on image of healing received from garden landscape.' *Journal of Agriculture Science, Tokyo University of Agriculture*, 48(3), 115–127. DOI: 110004027440

Akita City Board of Education. (2010) *Meisho Jyoshitei Hozon Kanri Keikakusho, (Jyoshitei Renovation Planning*, Akita, Akita City Board of Education.

Bullen, R. (2016) 'Chinese sources in the Japanese tea garden,' *Studies in the History of Gardens & Designed Landscapes*, 36(1), 5–16.

Collinge, J. (2021) Personal Communication 7 November.

Cooper, D. (2006). *A Philosophy of Gardens*, Oxford: Oxford University Press.

Daigoji (n.d.) www.daigoji.or.jp/grounds/sanboin.html (accessed 12 September 2021).

Daisen City Board of Education. (2010) *Kuni shitei keikatusho Kyu Ikedashi Teien Keikakusho (Ikeda family Renovation Planning)*, Daisen, Daisen City Board of Education.

Goto, S., Gianfagia, T. J., Munafo, J. P., Fujii, E., Shen, X., Sun, M., Shi, B. E., Liu, C., Hamano, H., and Herrup, K. (2016) 'The power of traditional design techniques: The effects of viewing a Japanese garden on individuals with cognitive impairment,' *HERD: Health Environments Research & Design Journal*, 10(4), 74–86. DOI:10.1177/1937586716680064

Habib, F., Nahibi, S., and Majedi, H. (2013) 'Japanese garden as a physical symbol of Japanese culture,' *International Journal of Architecture and Urban Development*, 3(4), 13–18.

Hamuro, R. (2017) *Sangetsuan Chakaiki*, Tokyo: Kodansha.

Hasegawa, T. (2021) *Shakaika Shiryoshu* [Social Science Reference], Tokyo: Kobun Shoten.

Hashimoto, T. (2016) 'Current issues on pesticides for medically-important pests,' *Trends in the Sciences*, 21(3), 72–76. DOI: 10.5363/tits.21.3_72

Hayakawa, M. (1973) *The Garden Art of Japan*, New York: Weatherhill.

Isoya, S. (1984) 'The beauty and the meaning of aging in Japanese Gardens,' *Journal of the Japanese Institute of Landscape Architects*, 48(5), 73–78. DOI: 10.5632/jila1934.48.5_73

Ito, S. (2002) *Bunmei to Shizen* [Culture and Nature], Tokyo: Tousuishobou.

Kanazawa Tourism Bureau (2020) *Kanazawa Kanko Hokokusho Tourism* [Tourism Report in Kanazawa], Kanazawa: Kanazawa Tourism Bureau.

Keane, M. P. (2003) 'Themes in the history of Japanese garden art,' *The Journal of Asian Studies*, 62(2), 623–624.

Kimura, T., Matsumoto, K., Okada, Y., Uchida, S., and Yamaoka, J. (2012) 'Psychophysiological effects by the appreciation of the garden and the art,' *Research Reports from the MOA Health Science Foundation*, 31–39. DOI: 40019547360

Kondo, D. (1985) 'The way of tea: symbolic analysis,' *Man*, 20(2):287–306.

Kuck, L. (1968) *The World of the Japanese Garden: From Chinese Origins to Modern Landscape Art*, New York: Weatherhill.

Miyagi Prefecture Board of Education (2021) Personal Communication, 8 September.

Mori. B. (1991) 'The tea ceremony: A transformed Japanese ritual,' *Gender and Society*, 5(1), 86–97.

Naka, T. (2009) *Nihon Teien no Mikata* [The way to look at Japanese Garden], Tokyo: Tokyo Bijyutsu.

Nonaka, N. (2008) 'The Japanese garden: The art of setting stones,' *SiteLINES: A Journal of Place*, 4(1), 5–8.

Patman, S. (2015) 'A new direction in garden history,' *Garden History*, 43(2), 273–283.

Sadler, A. (1962) *Cha-No-Yu: The Japanese Tea Ceremony*, Rutland: Charles E. Tuttle Company.

Sen, S. (1990) *Chado the Japanese Way of Tea*, Kyoto: Tankosha.

Stauskis, G. (2011) 'Japanese gardens outside of Japan: From the export of art to the art of export,' *Journal of Architecture and Urbanism, Vilnius Gediminas* Technical University, Lituania, 35(3):212–221.

Suzuki, M. (2007) *Japanese Gardens Outside of Japan*, Tokyo: Japanese Institute of Landscape Architecture.

Suzuki, M., Kurita, K., and Asou, M. (1994) 'Chinichika oubejin no nihonteien ni taisuru ninshiki to ime-ji ni kansuru chosa kenkyu [A study of the recognition and images of Japanese gardens in the minds of westerners familiar with Japanese culture],' *Journal of the Japanese Institute of Japanese Architecture*, 58(5), 5–8. DOI: 10.5632/jila.58.5_5

Tschumi, C. (2007) *Mirei Shigemori- Rebel in the Garden: Modern Japanese Landscape Architecture*, Basel Switzerland: Birkhauser Architecture.

Turay, M. D. (2001) 'The social significance of the Japanese tea ceremony,' *Japan Studies Association Journal*, 3, 49–64.

Wainwright, S. H. (2010) *Beauty in Japan*, London: Routledge.

Watanabe, M. (2005) *Seisho Bunka tono Setten* [Connection with Bible], Tokyo: Shinkyo Publisher.

Zhou, H. (2013) 'A Study on the space composition of borrowed scenery garden of Japan,' *Journal of Architecture and Planning*, 78(689), 1659–1666.

7

LITERATURE

Zeami Motokiyo (1363–1443) described flowers as enigma: 'flower kept in secrecy, it is the flower; if not concealed, it is not the flower (*hisureba hananari, hisezuha hana narubekarazu*)' (Morinaga 2005: 45). Flowers are not only used for *ikebana* arrangement, but also in literary works as well to express a person's inner feelings, spirits, beliefs, and messages to others. In this chapter, readers will explore how the flower has been used in various kinds of literature from antiquity until modern times in *The Kokin Wakashū*, *The Tale of Genji*, *The Pillow Book*, *Takekurabe*, as well as through the language of flowers (*hanakotoba*). At the end of this chapter, Lee Friederich also explores how women's experiences are evoked through flowers in both *The Tale of Genji* as well as a related modern work, *Onnamen (Masks)* by Enchi Fumiko.

The *Kojiki*

Composed by Ono Yasumaro at the request of Empress Genmei in the early eighth century, the *Kojiki* is a chronicle of stories in Japanese history. Numerous spirits appear, including Izanagi no Mikoto, Amaterasu Omikami, and Susanoo no Mikoto in relation to Yaoyorozu no Kami in the belief that spirits dwell in natural objects. These natural objects have often been related to plants and flowers. Tanaka (2004) states that 82 kinds of plants and flowers are described in the *Kojiki* and that they may bear one or more of these characteristics: appearing as plants themselves, as materials of an object or instrument, or as part of the name of persons, spirits, or places.

DOI: 10.4324/9781003248682-8

The Manyoshū *and* The Kokin Wakashū

The Manyoshū is the oldest collection of Japanese poems (*waka*), compiled during the Nara period around the same time as the *Kojiki*; there are approximately 4,500 poems in the collection and half of them are anonymous (Cushman *et al.* 2012, Mitsuru 1984). Cushman *et al.* (2012) states that around 1,650 poems are related to flowers and plants, with 150 different species and around 50 different flowers mentioned. Some poems simply read about plants themselves to express the seasonal changes; others poems used plants to express their feelings.

> *Hitorinomi,*
> *Mireba koshimi,*
> *Kamnabi no Yama no Momijiba taorikerikimi.*
> Alone on the mountains did I stand,
> And thought, as I stood, of my absent dear;
> And standing plucked with twitching hand
> Leaves of the fading year. Unknown author
> *Okada 1934: 470*

Okada (1934) states that this poem expresses the person's sense of loneliness through autumn leaves. The most mentioned flower was the Japanese clover, which is present in 141 poems.

> *Wagayado no*
> *Hitomurahagi wo*
> *Omouko ni misezu hotohoto chirashitsurukamo.*
> The inn's bush clovers
> Shown not to my precious love
> May be utterly scattered
> (Otomo-no-Yakamochi Volume8 No.1565).
> *Collinge 2021*

Japanese apricot (*ume*) and citrus (*tachibana*) were the second and third most commonly used flowers: 118 poems, 68 poems respectively (Figure 7.1).

> Wagasono ni
> *Ume no hanachiru hisakata no*
> *Ame yori yuki no nagare kurukamo.*
> In my garden,
> Plum blossoms scatter
> From heaven's rain
> Snow drifts down
> (Otomo-no-Tabito Volume5 No.822)
> *Collinge 2021*

FIGURE 7.1 Japanese apricot

Source: Photograph by the author.

On the other hand, *The Kokin Wakashu* is recognized as the first imperial anthology ordered by Emperor Daigo. It was compiled during the early Heian period mainly by Ki no Tsurayuki. There are around 1,100 poems, and 324 of which are related to flowers and plants. Among them, 39 different species are used, with the cherry blossom as the most commonly used flower, appearing in 61 poems. The second most common plant appearing in the compilation is the Japanese maple in 40 poems, followed by the apricot in 28 poems respectively (Takada 2009).

> *Yononaka ni*
> *taete sakura no nakariseba*
> *haru no kokoro ha nodokekaramashi*
> In the world,
> Spring hearts would be better off
> If they never knew cherry blossoms.
> Ariwara-no-Narihira (No.53)
> *Collinge 2021*

The Manyoshu were read by people from different backgrounds, from lower to upper class: guard men (*sakimori*), street performers, farmers, and aristocrats. About half of the poems are anonymous. In comparison, *The Kokin Wakashu* was read only by the upper class. Wild flowers were preferred in *The Manyoshu*, whereas garden flowers were more popular in *The Kokin Wakashu*. Cherry blossoms, wisteria, pink flowers, azeria, and camellias were also popular in *The Manyoshu*. However, the Japanese apricot, which came from China, was popular in *Manyoshū* due to the fact that the Japanese at that time appreciated Chinese culture more than their own Japanese culture. The cherry blossom became popular during the Heian period when *The Kokin Wakashu* was compiled (Mitsuru 1984) (Figure 7.2).

The Tale of Genji

The Tale of Genji written by Murasaki Shikibu (c. 973 or 978 - c.1014 or 1031) in the early eleventh century uses numerous flowers and plants for chapter names and to portray a character's feelings (Koga 1964). Murasaki Shikibu worked as a court lady. She wrote about the love story of Hikaru Genji, the son of an ancient Japanese emperor. Shikibu gave nicknames related to plants to each female character. Koga (1964) maintains that nicknames help readers to remember the characters. Murasaki no Ue (Lady Murasaki) was described as the ideal woman for Hikaru Genji because the color of the murasaki flower, purple, resembles the Lithospermum erythrorhizon, the purple promwell flower. Purple was perceived as the most superior and rich color; thus, it was implied that Murasaki no Ue was a noble and suitable person for him (Koga 1964).

FIGURE 7.2 Cherry blossom

Source: Photograph by the author.

This novel uses flowers as metaphors for the characters and their emotions. The moonflower (*yūgao*) was selected as the name of the female character who was not from the upper class, but would express a humble beauty. Koga (1964) asserts that morning glory, which blooms on the vine (*tsuru*) *and* does not have a straight stem, was selected as a suitable plant to express Genji's discrete love affairs with Yūgao.

On the other hand, blooming from a straight stem at a rather noticeable size, the white chrysanthemum, which grew in a character's garden, was described as the expression of a formal marriage proposal in this novel (Takashima 2006).

The Pillow Book

The Pillow Book (Makuro no Soshi) (1002) was written by Sei Shonagon (c. 966–1017 or 1025) during the Heian period. Shonagon had many opinions about flowers: if it is a flower tree, the Japanese apricot is the best whether the flower color is a faded one or strong. The cherry blossom petals should be big and have a dark color. It should be blooming on thin branches. Figure 7.3 is a comic describing *The Pillow Book* in modern Japanese. It is interesting to see that these

FIGURE 7.3 *The Pillow Book manga* for students

Source: © Gakken Plus.

works of classical literature, *The Kojiki*, *The Manyoshu*, *The Kokin Wakashu*, and *The Pillow Book* are studied as part of the syllabus for compulsory education in Japan. Students still learn how to admire flowers and plants, feel seasonal changes, and express their emotions using nature (Ishida 1979).

Modern literature

Higuchi Ichiyō (1872–1896), a well-known female writer, wrote *Takekurabe* (Child's Play) (1896), a novella depicting the story of Midori's life living in Yoshiwara, the red-night district while her elder sister was a popular courtesan. Ichiyō used a paper narcissus to express the impossible love from Nobuyuki, who was going to school to become a monk, for Midori, who is likely to become a prostitute herself (Figure 7.4). Kimata (1998) suggests that the daffodil (narcissus) implies a sense of purity. However, it was an artificial flower, which may also imply impurity in relation to her future career. It also has been interpreted that the artificial narcissus was intended to represent a wish for eternal purity (Kimata 1998). Was this wish for eternal purity of love the wish of the character, or merely a narrative device employed by the author to represent the metaphor of everlasting purity? (Collinge 2021) Kimata further comments that this love story was also related to Ichiyo's own unrequited love for her editor.

Brought up as the son of a wealthy landowner in Aomori, Dazai Osamu (1909–1948) wrote numerous novels. In the novel *Fugaku Hyakkei* (One Hundred Views of Mount Fuji) (1939), he described his thoughts and emotions in relation to Mt. Fuji and flowers. He had a negative impression of Mt. Fuji from the fact that it

FIGURE 7.4 Daffodil

Source: Photograph by the author.

was always admired by most Japanese; he was a person who rarely liked what the majority favored. Dazai described his impression of his future wife as a white water lily that represented purity. On another occasion, he wrote, 'The evening primrose suits Mt. Fuji well.' The flower of the evening primrose (*tsukimiso*) has a humble appearance and blooms only in the evening (Matsuda 1978). Matsuda (1978) has argued that this humble-looking flower represented Osamu himself, expressing his writing skills as still not the utmost compared with Mt. Fuji. There is another scene in which Dazai describes the view of girls wearing red coats visiting Mt. Fuji with small anemone flowers in the background. While Dazai did not state the color of these flowers, Matsuda commented that the anemone was highlighting the young girls' pure spirit. At the end of the scene, Dazai's description of Mt. Fuji is like the Chinese lantern plant (*hoozuki*). Some critics interpret this plant's inclusion as a symbol of a false image expressing Dazai's belief that Mt. Fuji was no longer his rival. Others interpret this plant as the toy for little toddlers, and that in the end, influenced by the fact that he had found his fiancée, he felt relaxed about Mt. Fuji.

Haiku

Haiku is a Japanese poem constructed with 17 syllables, divided into 3 syllabic sections following the 5, 7, 5, syllable pattern. This style was created in the Edo period and is still commonly made to express our feelings. It must also contain *kigo* (seasonal phrases) (Barnhill 2012). Flowers and plants are used for *kigo*. Table 7.1 shows some examples. There are several *kigo* related to colored leaves which can be used to express the delicate seasonal changes or feelings.

The language of flowers (*hanakotoba*)

Flowers convey messages, and the custom of passing messages through flowers still exists in Japan today. This custom became known as Floriography (*hanakotoba*) after the Meiji period. The custom of conveying messages through flowers became popular in the fifteenth century in Turkey among harem members as a communication tool. Later, it spread to Europe and came to Japan in the early Meiji period (Mancoff 1996). Hanakotoba is often seen in newspapers and magazines in Japan. The language of flowers is influenced by myths and legends, characteristics, histories or customs, and colors and numbers (Mancoff 1996). Therefore, flowers sometimes convey different messages depending on the country. For instance, while cherry blossoms represent graceful women in Japan, they mean good education in the UK (Kurashima 2019) (Table 7.2).

Cherry blossom (*sakura*)

Yosano Akiko (1878–1942), a prominent female poet also adored cherry blossoms (*sakura*) and named her final poem book after *sakura*, *Hakuoshu*. Although *sakura*

TABLE 7.1 *Kigo*

Name	Japanese name	Season
Wisteria	*fuji*	late spring
Iris	*ayame* or *shōbu*	early summer (May)
Orange blossoms	*mikan*	early summer
Lily	*yuri*	mid summer
Water lily	*suiren*	mid and late summer
Lotus flower	*hasu* or *hachisu*	late summer
Sunflower	*himawari*	late summer
First colored leaves	*hatsu-momiji* or *hatsu-momijiba*	mid autumn
Leaves turning color	*usumomiji*	mid autumn
Colored leaves	*momiji* or *kōyō*	late autumn (October)
Leaves start to fall	*momiji hatsu chiru*	late autumn
Shining leaves	*teriha*	late autumn
Tachibana orange	*tachibana*	late autumn
Holly	*hiiragi*	early winter
Dry leaves	*kareha*	early, mid, and late winter
Fallen leaves	*ochiba*	early, mid, and late winter
Winter chrysanthemum	*fuyugiku* or *kangiku*	early, mid, and late winter
Daffodil (Narcissus)	*suisen*	late winter
Ornamental kale	*habotan*	late winter

Source: Barnhill 2012.

TABLE 7.2 Word of flowers

Flower	Japanese name	Japanese name meaning
Peony	*botan*	Bravery
Poppy	*keshi*	Success
Red Rose	*aka bara*	Love/in Love
White Rose	*shiro bara*	Innocence
Tulip	*chūrippu*	One-sided love
Zinnia	*hyakunichiso*	Loyalty
Daffodil	*suisen*	self-love, egotism, 'Come back to me.' (yellow)
Hydrangea	*ajisai*	Flirtation, family, harmonious

Source: Kurashima 2019.

has not been recognized as the national symbol of Japan, along with the chrysanthemum, it has been used for various metaphors for Japanese lives and occasions (Ohnuki-Tierney 2002) (Figure 7.5). *Sakura chiru*, or 'fallen sakura,' implies that a person has failed an exam, could not accomplish their wishes, or has died. On the other hand, *Sakura saku*, 'sakura blooms,' implies that if a person passes the exam, their wishes will come true. *Sakura* was also used as the metaphor of transient

FIGURE 7.5 *Sakura* in the mountain

Source: Photograph by the author.

hakanai (short lived; momentary) life of the *kamikaze* pilots in Japan during the Second World War (Ohnuki-Tierney 2002). Ohnuki-Tierney (2002) argues that well educated *kamikaze* pilots perceived the cherry blossom not in militaristic terms, but as a symbol of their beautiful transient life; it appears that the pilots and others understood their tragic fate, but used the beautiful short-blooming *sakura* to represent their lives as a way of consoling themselves about their painful realities.

In Japanese society, delicate feelings are not openly displayed in public, which is illustrative of the *honne/tatemae* (real feeling/public behavior) and *haji* (embarrassment or shame) concepts in Japanese society (Hendry 2019, Chiba 2010). From a young age, Japanese are not encouraged to express their real feelings or emotions in public. If this is not possible, sometimes it is perceived as an embarrassing matter. Due to this context, it appears that flowers and plants have been recognized as the appropriate tools for expressing one's passionate emotions and feelings to another person. This way of expression has lasted a long time, and even today, the practice can be seen in pop culture, including in popular Japanese pop music. The song *Hanaga Saku* (The flower will bloom again) was made in memory of the 2011 earthquake in the Tohoku area. The song expressed the intense hope that Tohoku would recover from the catastrophic natural disaster, much like flowers that bloom once more after bad weather. The strong sentiment of hope is still present in Japanese society today, and the song continues to be sung by many generations across Japan.

Research questions

1. Why is it important for educators that children learn to appreciate flowers and nature through literature at a young age? Does the appreciation of flowers in literature influence the children's upbringing?
2. The *Manyoshu* was read by people from different backgrounds, while *The Kokin Wakashu* was read only by the upper class. What was the difference? Was it hard for those who lacked the knowledge to read *The Kokin Wakashu*? Or, was it deliberately withheld from the lower classes as a means of symbolically establishing and maintaining the positions within the social casts of that period?
3. Is it unique to Japanese culture to use flowers as metaphors for the characters in narrative texts and their emotions? Or, are there other cultures that use flowers to depict characters or their emotions?
4. How were the feelings and thoughts towards flowers in the past spread among people? Was this actively encouraged and deliberate?
5. Japanese have associated their feelings with flowers, but why flowers when it could have been rocks, grasses or animals?
6. Many Japanese songs have flower names in their titles and lyrics. How is it similar or different to the use the names of flowers in literature?

Conclusion

This chapter highlighted and investigated the frequent use of flowers and plants in Japanese poems, novels, and in Japanese poetry and prose. Japanese literature commonly uses flowers and plants to subtly suggest a sense of beauty that has come from a person or an object, as well as the beauty of the changing seasons, or to provide social meaning. Poets in the *Manyoshu* sensed the delicate seasonal changes and portrayed certain feelings through imagery and metaphors pertaining to plants and flowers. It still appears to be common for the Japanese to sense the seasonal changes and acknowledge their emotions through the use of the flowers and plants in twenty-first century Japan. Even today, since expressing one's passionate emotions and feelings to another person is not common within the concepts of *honne* and *tatemae* in Japanese society, flowers and flower arrangements might be forever appreciated as methods of conveying our innermost feelings.

References and further reading

Barnhill, D. (ed.). (2012) *Basho's Haiku: Selected Poems of Matsuo Basho*, New York: SUNY Press.

Chiba, K. (2010) *Japanese Women, Class and the Tea Ceremony: The Voices of Tea Practitioners in Northern Japan*, London: Routledge.

Collinge, J. (2021) Personal Communication November 30.

Cushman, S., Ramazani, J., Rouzer, S. P., and Cavanagh, C. (2012) *The Princeton Encyclopedia of Poetry and Poetics: (4th ed.)*, Princeton: Princeton University Press.

Dasai, O. ([1939] 2012) *Fugaku Hyakkei* [One Hundred Views of Mt. Fuji], Tokyo: Aozora bunko.

Hendry, J. (2019) *Understanding Japanese Society*, London: Routledge.

Higuchi, I. ([1896] 2004) *Takekurabe* [Child's Play or Growing Up] (R. Matsuura, S. Fujisawa, H. Shinohara, A. Itsuji, & K. Abe, trans), Tokyo: Kawade shobo shinsha.

Ishida, J. (1979) *Makura no Soshi* [The Pillow Book], Tokyo, Kadokawa.

Kimata, T. (1998) 'How to read the artificial narcissus' [Suisen no tsukuribana wo douyomuka], *Japanese Literature Association*, 47(11), 12–21. DOI: 10.20620/nihonbungaku.47.11_12

Kassim, F. (2014) 'Status of women in the heian period: A study of the literary works of Murasaki Shikibu and Sei Shonagon,' *WILAYAH: The International Journal of East Asian Studies*, 3(1), 111–124.

Koga, M. (1964) 'The tale of Genji and plants: The effects of the use of plants' [Genji monogatari to shokubutsu: Shokubutsu shiyou no kouka], *Kumamoto Women's University Kokubun Danwa kai*, 10, 21–42.

Kurashima, A. (2019) *Hana no Kotoba Jiten* [Word of Flower], Tokyo: Kodansha.

Mancoff, D. N. (1996) 'The language of flowers: A history,' *Victorian Studies*, 39(4), 604–605.

Matsuda, T. (1978) 'Flowers in Dazai's literature' [Dazai bungaku ni okeru hana], *Osaka Shoin Women's University Journal*, 16, 84–90.

Mitsuru, S. (1984) *Flowers of Manyo* [Manyo no hana], Tokyo: Yuzankaku publish.

Morinaga, M. (2005) *Secrecy in Japanese Arts: 7 Secret Transmission' as a Mode of Knowledge*, London: Palgrave Macmillan.

Ohnuki-Tierney, E. (2002) *Kamikaze, Cherry Blossoms, and Nationalisms: The Militarization of Aesthetics in Japanese History*, Chicago: University of Chicago Press.

Okada, T. (1934) 'English translations of the manyoshu,' *Studies in English Literature*, 14(4), 468–481. DOI: 10.20759/elsjp.14.4_468

Osawa, K. (2013) 'Actions against plants in the manyoshu and imperial anthology of Japanese poetry of the heian period' [Manyoshu oyobi heian-ki no kyokusen waka shū ni miru shokubutsu ni taisuru koui], *Journal of the Japanese Society of Revegetation Technology*, 39(1), 74–79. DOI: 10.7211/jjsrt.39.74

Takada, H. (2009) *Kokin Wakashu*, Tokyo: Kadokawa Sochia.

Takashima, K. (2006) *Genji monogatari shokubutsu kou (1)* [A Study of Plants in the Tale of Genji (Vol. 1)], Tokyo: Kokken Shousho.

Tanaka, C. (2004) 'Plants in "Kojiki": On the Relation between the Number of Appearance and the Story ["Kojiki" kisai no shokubutsu: Toujou-sū to monogatari no kanren ni tsuite],' *Konan Women's University Graduate Studies in Literature and Culture* (2), 17–25.

IN THE NIGHT GARDEN: TAMAKAZURA REVISITED IN ENCHI FUMIKO'S MODE RN-DAY TALE OF FAMILY TRAUMA

Lee Friederich

The flowers we encounter in Murasaki Shikibu's The Tale of Genji are as diverse and evocative as the women they represent. Genji, not even the narrator of this work, plays a strong role in how women are named, gifting his greatest loves with the name of a flower or plant in her vicinity. Flowers blossom throughout the monogatari in traditional waka poems exchanged by lovers that allow not only the expression of love, but also memorialize love lost. Flowers also show the strong associations Genji makes among his great loves as one 'flower' unfolds into another. For instance, when he expresses his desire to make the young girl he will later call Murasaki his own 'little wild plant,' Genji is also evoking his illicit passion and continued longing for his step-mother, Lady Fujitsubo (Murasaki 1008/2001: 100) by associating both with the color purple: wisteria in the case of Fujitsubo and in Murasaki's name, the roots of the plant that produces a deep purple dye 'associated with love' (Tyler 2001: 100). Unsurprisingly, Genji's mother, Lady Kiritsubo, who died when Genji was only three, is also associated with a purple flower: Kiri means paulownia or 'princess tree,' which also bears purple flowers. Never needing to be re-planted, the paulownia regenerates spontaneously, a process that helps us understand the ways in which Kiritsubo's maternal powers inform so many of Genji's relationships ('Regeneration of a tree,' 2020). Flowers take on a wholly different feeling for the women they represent when we examine how women describe themselves in the waka they exchange with Genji. Briefly exploring Tamakazura's expression of abandonment, gloom, anxiety, and vulnerability in the face of Genji's 'fatherly' love, a dreadful love she recognizes as a projection of his obsession for her long-deceased mother Yūgao, this essay takes up Enchi Fumiko's transformation of Tamakazura in her modern day tale of possession, Masks.

Genji not linking a wife or a lover with a flower more than likely signifies a problem in the relationship. Not affiliated with a flower when we first meet her, Lady Rokujū's sense of exclusion motivates the monogatari's plot: Rokujū's all-consuming love turns to an all-consuming jealousy that results in her spirit possession of various women, beginning with Tamakazura's mother, Yūgao. Capturing Genji's heart with her delicate charm, Yūgao, the 'twilight beauty,' is named for a white flower with a 'bright green vine' that is 'clambering merrily over a…board fence' (Murasaki 1008/2001: 55). Offering Genji a fan on which to keep the flower he will eventually name her for, Yūgao is nonetheless intent on withholding her identity from him, which only adds to Genji's attraction to her. Genji resumes his search for Yūgao again when Tamakazura, who was an infant when her mother died, turns up in his life as suddenly as her mother had vanished.

Genji views Tamakazura as the woman he wishes her to be: her mother, Yūgao, more so than seeing her as the woman Tamakazura has become. Replicating his relationship with Murasaki, he 'adopts' Tamakazura, insisting he wishes to play a fatherly role in her life. Old enough to be her father, Genji expresses his 'lasting' bonds to Tamakazura in the image of 'Mishima reeds,' a reference to poetic lines that epitomize the '"connection" between one person and another' (Tyler 2001: 422). Tamakazura picks up on the image of the reeds, not as a karmic connection with Genji, but with her dead mother. Instead of emphasizing Genji's bonds with her mother, we see her own emotional concerns as a 'deep root' of sorrow, knowing neither her mother nor father. Also identifying herself as her mother's daughter in her piercing, self-referential use of natural images, Tamakazura pointedly replies to Genji,

'Whence does her stem spring, this unhappy reed whose work is so very small, that she has struck such deep root into the sorrows life brings?' (Murasaki 1008/2001: 422)

The poem betrays Tamakazura's profound 'distress' at receiving a message from a man other than her own father, 'a man she does not even know' (Murasaki 1008/2001: 422), a distress that will cascade throughout the Tamazakura chapters that occupy some 100 pages in the last part of the first half of the monogatari.

Early on in the famous Fireflies chapter, a chapter that Enichi reconfigures in Masks, we sense that Genji's enjoyment at dangling Tamakazura before would-be suitors is not only a source of 'anguish and perplexity' for her, but makes her long for the other parent who could protect her: She 'grieved afresh for her mother and bitterly regretted her loss' (Murasaki 1008/2001: 435). It is against this backdrop of despair that her face is revealed by the light of fireflies for the benefit of the 'firefly' prince, Tamakazura's latest suitor, Genji's brother, Hotaru. A voyeur to this scene, Genji scents the air with incense as he positions himself close by to take it all in, as the narrator puts it, in 'hateful and unfatherly perverseness' (Murasaki 1008/2001: 436). Signaling Tamakazura's emotional isolation in her relationship with Genji, others construe the prank as the highlight of a successful evening, ironically, as 'the motherly care His Grace had given their mistress that evening' (Murasaki 1008/2001: 457). Eventually, Genji recognizes he has no choice but to hand Tamakazura over to her actual father, Tō no Chūjō, so that 'the truth' of his relationship with the woman he has also claimed as his daughter would not leave him with 'a lasting and unfortunate reputation as a master of devious plots' (Murasaki 1008/2001: 502).

*

Gessel (1988) proposes that Enchi Fumiko may have viewed her novels *Masks, The Waiting Years*, and *A Tale as False Fortunes* as a trilogy that explores the perspectives of various female protagonists in the Genji Monogatari: Lady Rokujō, Lady Aoi, and Lady Murasaki respectively. Finishing her translation of *The Tale of Genji* into modern Japanese in 1973 after penning these three works, Enchi makes strong use of flowers in *Masks* to reference some of the women we encounter in *The Tale of Genji*, but with a hauntingly different effect. Unlike Lady Rokujō, whose possessions are subconscious, Mieko knows what she is doing, though she so often conceals her thoughts with a mask-like expression: 'She's like the face of a Nō mask, wrapped in her own secrets,' as her daughter-in-law Yasuko puts it (Enchi 1958: 30). At the same time, Mieko's actions extend beyond her responses to her experiences as a woman whose life spans the Meiji through the Showa eras. Seen as a meditation on 'women's evil,' Masks reveals both Mieko's unforgivable plot in which she sacrifices her only daughter, while also suggesting her awareness of the origins of her evil as a product of patriarchy. As Mieko reveals in the conclusion of her essay on Lady Rokujō 'An Account of the Shrine in the Fields,' which functions as a kind of prologue to her plot, she is aware that women's evil cannot be separated from the male world in which she is both 'an object of man's eternal love' and 'the object of his eternal fear' (Enchi 1958: 57). Interestingly, flowers and the secrets that they hold, become an especially pregnant metaphor to reflect the complexity of Mieko's heart, mind, and actions.

Recognizing that Genji so often associates his greatest loves with flowers, Mieko remarks in her essay that Lady Rokujō had not been properly introduced in the monogatari because Genji had not given her a flower sobriquet. Pointing out that women such the Evening Faces Lady (Yūgao) had been named for a flower to commemorate their 'distinctive beauty,' Mieko also implies that it might be difficult for a flower to emblemize Lady Rokujō's qualities as a woman, for she is wholly unlike the women named for flowers, who, in her words, 'dissolve their whole beings in the anguish of forgiving men, and thereby create an image of eternal love and beauty in the hearts of the men they love' (Enchi 1958: 48–49, 52). Rather than to compare her to a flower, Mieko compares Lady Rokujō to the Noh mask Ryō no onna, as 'one who chafes at her inability to sublimate her strong ego in deference to any man...' (Enchi 1958: 52).

While Lady Rokujō confronts her victims physically through spirit possession without awareness of what her wandering spirit is doing, Mieko knowingly encapsulates her own refusal to 'defer...to any man' through action cloaked in silence. Striking back at her husband, Masatsugu, for his infidelity through an affair of her own that produces her twins Akio and Harume, Mieko thwarts his bloodline without telling him that his children in fact were those of another man. Betraying her deep empathy for Rokujō, who, as Mieko writes in her essay, 'can carry out her will only by forcing it upon others…through the possessive capacity of her spirit' (52), Mieko, confronted with the early death of her son, schemes to find a way to carry this new bloodline into the next generation. Herein lies her plot to have Akio's intellectually disabled twin impregnated, thereby perpetuating her lover's bloodline even further. We might even say that Genji's misuse of his parental privilege in his dealings with his 'daughter' Tamakazura pales somewhat in comparison when Mieko puts her daughter Harume up for rape: Harume dies after giving birth to the child conceived in Mieko's plot. While Mieko cannot be forgiven for what she does, her desperate actions can be understood as a response to Masatsugu's unwillingness to let her leave his family when she learned of his long-term affair with his maid, the person responsible for the death of the child she was carrying.

There is seemingly no end to the cruelty mothers and their offspring face in *Masks*: Forced to undergo two abortions to end her pregnancies by Masatsugu, his longtime maid and lover, Aguri takes her revenge against Mieko and her unborn child. Knowing that Mieko's *kimono* would catch on the nail she planted on the stairwell, Aguri was all but assured that she would kill one or the other: Mieko or the fetus in her womb. Making her husband the object of her revenge rather than Aguri, we might view Mieko's actions in the same way Bargen (1997) views Rokujō's alliance with Lady Aoi, who ultimately dies in their joint effort to show their pain to Genji through Rokujō's possession of her. Making her own daughter a surrogate for her revenge, Mieko shows her colors as a woman born into the Meiji era, a society steeped in the power of blood relationships. Like Genji, she willfully asserts her authority as a parent to her advantage. Mieko understands and yet cannot find her way out of this patriarchal trap of hatred: In a vision she has after orchestrating her daughter's rape, she associates herself with the 'ancient goddess,' Izanami, whose 'love she had borne' her husband Izanagi 'transformed utterly into blinding hatred' (Enchi 1958: 127). Unable to conceive alternative ways in which to survive, Mieko recognizes that her own 'passion for revenge' is part of a larger 'obsession that becomes an endless river of blood, flowing on from generation to generation' (127).

Taking full advantage of her only living child Harume, Mieko also manipulates Akio's widow, her daughter-in-law Yasuko, who pushes Harume into the sexual encounter with Yasuko's current lover, Ibuki. 'Possessing' Yasuko, who acknowledges that like Rokujō, Mieko 'has a power to move events in any direction she pleases.' If Yasuko knows anything about Mieko, she knows that 'the secrets in her mind are like flowers in a garden at nighttime, filling the darkness with perfume' (Enchi, 1959, pp. 30, 32). As Yasuko suggests, throughout her plot of 'possession,' Mieko's intentions remain hidden, even from Yasuko, who nonetheless emerges from Mieko's plot as a co-conspirator, so strong is their bond of grief over the death of Mieko's son, Yasuko's husband. Remaining impassive before others, Mieko is visibly shaken about what she is doing to her daughter. Alone in her room during her daughter's rape and conception, she 'sees' her daughters 'eyes wide with alarm' and hears her 'inaudible cry,' which she describes as 'an unearthly, astonishing voice' (Enchi, 1958, p. 106). Unlike Rokujō, she is completely aware, in every bone in her body, of what she is doing and the pain she is inflicting on Harume to carry out her plot 'that only women could commit,' as Yasuko puts it (Enchi 1958: 126).

While Mieko takes on the persona or at least the fragrance of a delicate flower – wordless, serene, and in the end, deadly toxic – we are never sure exactly what Mieko has conveyed of her plot to Yasuko; after carrying out Harume's rape though, Yasuko claims to feel even a greater bond to Mieko. Acknowledging that she thinks it is 'inhuman…to make a woman like her, with so many physical and mental handicaps risk childbirth,' Yasuko exclaims that their crime means more to her 'than the love of any man' (Enchi 1959: 126). In the end, neither Ibuki nor Mikame, both of whom are in love with Yasuko, can fathom Mieko's motivation for duping Ibuki into an affair with Harume. Even though they have surmised her admiration for Lady Rokujō through their discovery of her essay, as well as the details of Mieko's miscarriage, based on their own undercover work into her background, try as they may, they cannot fathom the 'secrets' beneath Mieko's calm exterior, which, as Yasuko claims, 'had all the fragrance of a garden at nighttime' (Enchi 1959: 133).

Significantly, Harume herself, whom Yasuko discusses with Ibuki earlier on in the novel, is first glimpsed physically, as 'a large white flower…magnificent in her isolation' at a Tale of Genji-themed firefly party that Mieko is hosting in honor of a colleague (Enchi 1959: 39). Seeing her through the eyes of Ibuki and Mikame, both of whom are vying for Yasuko's love, we see the way in which Harume has likely been 'framed' for the men to happen upon in this enchanting night garden: 'the figure of a woman seated alone in the arbor, gazing toward them. Light from the stone lantern cast a hazy circle on the umbrella-shaped pines. The woman's face, beside the lantern, was arrestingly beautiful…like a large white flower bathed in light, magnificent in her isolation' (Enchi 1959: 39). What the men do not know is that they, too, are included within this frame. Joking before encountering Harume that perhaps the elderly professor for whom the party was being hosted may be playing the role of Prince Hotaru for Yasuko's Tamakazura, it is easy to see that Mikame comes closest to playing the role of Hotaru as he quickens his pace to get closer to Harume, planted into the scene as Tamakazura. Then, as though appearing 'seemingly out of nowhere' and standing 'before the arbor as though waiting for them,' Yasuko blocks Mikame's path before he and Ibuki could reach her sister-in-law. A seasoned reader of the work might see this encounter as one of the early, entirely orchestrated scenarios leading to Harume's rape. While Yasuko can be seen potentially protecting Harume in this scene, when we reflect back on this scene after Yasuko has pushed Harume into the arms of her own lover, we might also feel the anguish of Tamakazura's isolation as she is exposed by Genji to Hotaru. Possessing the placid beauty of 'the flower,' Harume must also be seen as bait with which Mieko and Yasuko draw these two unsuspecting men, one of whom will father the child who will carry Mieko's legacy into the future, into their plot. We can also sense in Harume's vulnerability Tamakazura's longing for her mother Yugaō, the 'twilight beauty,' who is also absent when her daughter needs her most.

References and further reading

Bargen, D. (1997) *A Woman's Weapon: Spirit Possession in The Tale of Genji*, Honolulu: University of Hawai'i Press.

Enchi, F. and Carpenter, J. (1983) *Masks*, New York, NY: Random House.

Gessel, Van C. (1988) 'The "medium" of fiction: Fumiko Enchi as narrator,' *World Literature Today*, 62(3): 380–385. DOI: 10.2307/40144284

Kano, A., and Fumiko, E. (2006) 'Enchi Fumiko's stormy days "arashi" and the drama of childbirth,' *Monumenta Nipponica*, 61(1), 59–91.

Murasaki, S., and Tyler, R. (2001) *The Tale of Genji*, New York, NY: Penguin Classics.

Paulownia Professional. (2020) Regeneration of a tree. https://paulownia.pro/en/paulownia/

8
GENDER AND CLASS

Previous chapters have explored how flowers have been used to express our feelings. It appears that flowers were used by and for both genders, but how about *ikebana*? The majority of *ikebana* practitioners are female, so it is common for *ikebana* to be interpreted as an art mainly for women. However, *ikebana* used to be only accessible by men, and only after the Meiji Period (1868–1912) was it officially made available to women. This chapter argues that *ikebana* contributes to the perpetuation of gender roles and social class hierarchies to a certain extent. It further examines that *ikebana* simultaneously provides a great sense of gender empowerment due to the opening of the *iemoto* position to women. Readers will also explore the motivation behind *ikebana* practitioners compared with *chadō* practitioners. In order to explore gender and class issues in *ikebana*, let us review some historical background first.

Ikebana as a subject for girls' schools

As discussed in the history chapter, the number of *ikebana* practitioners increased during the Edo period (1603–1868). Much like other traditional Japanese arts such as *chadō* and *noh*, *ikebana* was practiced mainly among the upper and upper-middle class males from *samurai* and merchant backgrounds (Ito 2017). Ikenobo, the largest and oldest of all *ikebana* schools, accepted only males of the elite classes and refused to instruct women and male commoners (Nishiyama 1962, Stalker 2018). However, from the middle of the Edo period, upper or upper-middle class women also started to engage in these art forms in their own private lessons. These customs led to the introduction of *ikebana* education in girls' schools from the Meiji period (1868–1912) onwards (Akitaken Kado Renmei 2007).

DOI: 10.4324/9781003248682-9

In relation to the notion of a rich country, strong military (*fukoku kyōhei*), the Meiji government encouraged women to be the ideal 'good wife, wise mother (*ryōsai kenbo*).' This ideology promoted the idea that women were responsible for creating a pleasant home environment to help nurture the family and protect the nation (Koyama 2012, Stalker 2018: 105). With the rise in popularity of this ideology, the Ministry of Education mandated Etiquette as a subject for girls' higher education (Akitaken Kado Renmei 2007). *Ikebana* and *chadō* along with sewing and home economics were selected as the subjects for girls in their higher education. Stalker (2018) also states that the support of the Meiji government in encouraging the teaching of girls arose from the desperate desire for Japan to be recognized in the USA and Europe, where flower arranging and tea parties were already regarded as feminine pursuits, as a developed nation (Jackson-Houlston 2006).

Atomi School was founded in 1875 as Japan's first private girl's school. *Chadō* and *ikebana* were taught as compulsory subjects along with Japanese, Chinese, mathematics, calligraphy, painting, and sewing (Kido 2007). The Atomi School greatly influenced Gakushuin, Keio Girls School, Seiken Girls School in Tokyo, Horikawa School in Kyoto, and other public schools, all of which also started *ikebana* classes. Fukuzawa Yukichi (1835–1901) commented on the importance of traditional arts education for girls, and in his book *Women's learning* (*Onna-daigaku*, 1899), encouraged families to let their daughters engage with these art forms for as long as they could. Supported by the well-recognized advocator, Fukuzawa, *ikebana* education for girls began to be adopted elsewhere in Japan (Ito 2017). Stalker (2018) argued that by the 1920s and early 1930s, *chadō* and *ikebana* were universally accepted as a significant part of girls' curriculum. However, it is significant to note that not every child in Japan went to elementary school, thus only a limited number of upper-middle and middle class girls were able to attend a girl's school (Sugimoto 2015). In Japan at that time, this was considered as optional education after completing compulsory education, whereas the mixed compulsory education started for those aged 6–12 later in the Meiji period (Andreesen and Gainer 2002).

From the Meiji period, *ikebana* began to be perceived as the subject to learn manners and etiquette due to the introduction to girl's education; it was regarded as suitable moral education for girls while the way of *samurai* (*bushido*) was taught to boys (Akitaken Kado Renmei 2007). Ito (2017) argues that *ikebana* style changed from a masculine to feminine style. While the large *ikebana rikka* style was popular, the size of *ikebana* became smaller. *Rikka* style was more detailed and had to follow detailed rules, whereas *Seika* style had simpler rules (Figures 8.1 and 8.2). Goldstein-Gidoni (2005) further commented that *ikebana* schools taught the simpler *ikebana* style to women since flower schools and the government's main purpose was to simply let women acquire discipline to become good wives and mothers, not to let them learn in-depth *ikebana* techniques. Moreover, *ikebana*

FIGURE 8.1 *Rikka*

Source: Photograph by the author.

was perceived as a subject related to uneconomic subjects in which only girls from the high society could engage (Akitaken Kado Renmei 2007).

Ikebana for war widows

After the Russo-Japanese War in 1904, Ikenobo made teaching positions available for women, especially for widows who had lost their husbands in the war. Skalker (2018) states that a considerable number of middle class women began teaching *ikebana* when many men left the specialized profession. Taira Ichiyo, a war widow, was one of the first female *ikebana* teachers in Japan. Taira had been learning *chadō* and *ikebana* since she was young. She had lost her husband in the Russo–Japanese war, and raised her seven children by becoming a *ikebana* teacher. She supported Ohara school and made this school successful (Skalker 2018). Teaching *ikebana* became an acceptable occupation, with some schools, including

FIGURE 8.2 *Seika*

Source: Photographed by the author.

Ohara school, making the training time clear and short, so that practitioners were able to receive qualifications faster than before.

Ikebana for bridal training

After the Second World War, the Japanese education system was reformed and became accessible to every social class (Sugimoto 2015). Yet, fearing that traditional culture might reignite nationalistic and feudalistic sentiment, many traditional arts and cultural values were re-questioned (McLelland 2012). *Terakoya no dan* is one example of the *kabuki* plays forbidden by GHQ (General Headquarters of the Supreme Commander for the Allied Powers), which was concerned that the play promoted absolute loyalty to the outdated feudalistic sentiment. As a result, the ministry of education ascribed less importance to teaching traditional culture (Kokubun 2007). *Ikebana* was mainly taught as an extracurricular activity, rather than an actual subject in schools.

Although traditional arts education was not promoted in compulsory education, the image of bridal training was upheld through the continued practice
of *ikebana* (McVeigh 2005). *Ikebana* was perceived as bridal training (*hanayome
shugyō*) for upper- and middle-class women and taught mainly at the *ikebana*
teachers' houses. It was common that women in their early twenties practiced
ikebana for a couple of years before their arranged marriage (Figure 8.3). *Ikebana*
was categorized as high culture for the upper- and middle-classes in Japan. It was
commonly understood that a lot of time and patience was needed to acquire and
understand the skills and philosophy required to perfect the art. In addition, it
also costs a substantial amount of money to become a licensed practitioner. By
stating that a person was engaging in *ikebana* practice was to affirm their social
class and status, *haku ga tsuku* (gold is attached). Stalker states that the *ikebana*
population reached its peak in the 1960s. Nishiyama estimated that there were
ten million *ikebana* practitioners in 1965, roughly 10 percent of the population
(Stalker 2018).

Stalker (2018) argues that *ikebana* appreciation among Japanese women was
also related to the popularity among foreign women during the occupation.
According to Stalker, 95 percent of foreign women in Japan tried *ikebana* during
this era; many became diligent students and over 10,000 became licensed teachers

FIGURE 8.3 Women getting married in 1950

Source: Photograph by a person unknown.

(Stalker 2018: 104). These interests by western women also contributed to the status that *ikebana* provided Japanese women.

Ikebana as status symbols for women?

So far, we have explored how *ikebana* has contributed to bridal training (*hanayome shugyō*); however, the concept of *hanayome shugyō* has died out. Today, middle class women in their twenties are more engaged in job hunting than considering finding an appropriate husband. Very few of the younger generation are engaged with *ikebana* (Agency of Cultural Affairs 2018). Yet, *ikebana* is still favored by women rather than men, and there are still practitioners actively engaged with *ikebana*. According to the Agency for Cultural Affairs, only 2.2 percent of men had practiced *ikebana*, while 35.7 percent of women answered that they had engaged in *ikebana* practice (Agency for Cultural Affairs 2018).

The following are interviews with Sawata-*san*, Tomoko-*san*, and Kodama-*san*, current *ikebana* practitioners and their family members who have shared their experiences, motivations, and understandings of *ikebana*. Akita prefecture encompasses 16 different *ikebana* schools, and there are around 500 *ikebana* practitioners, 99 percent of whom are women, the majority of whom are over 70 years old (Terada 2021). Sawata-*san*, originally from Sendai, in her 40s, shared her thoughts about her aunts who have been engaged with *ikebana* for over 30 years.

> My aunt is busy with *ikebana, chadō*, and many other hobbies. Her husbands worked for relatively big trading companies. She never worked, and did not have any kids. She really enjoys practicing many kinds of activities. She recently lost her husband, so I am busy supporting her. She calls me every day for a couple hours.

She further comments about *ikebana* activities:

> Yeah, these days we do not hear that our friends are practicing *ikebana*, right? They are only my aunt's generation. Nobody does it. It costs money, and we need time to do it. The *ikebana* image is like somebody who can afford *ikebana*, I think. My aunt can afford it I guess, her husband works for a relatively big company. She does not work and spends all her time on her hobbies. Do you know how much money she spends on the stage for her singing? 3 million yen!

Sawata-*san* further continues:

> You see around our *mamatomo* (mother's friends), most of them are working, they are just busy maintaining work and housework. They do not have

time for hobbies. If there is free time, they tend to do something related to their health benefits, like yoga, or going to the gym.

Tomoko-*san* in her 70s, used to practice *ikebana* when she was younger and shared her life stories and image of *ikebana*:

> I was born in Oodate, the north part of Akita prefecture. I was born as the daughter of a Japanese restaurant owner. My mother sent me to many practices, typically *chadō*, *ikebana*, calligraphy, and so on. For me, *ikebana* is the image of *KateiGaho*, you know the magazine for MADAMU, women with elegance, time, and money. Those who practice *ikebana* are people who have enough finances and time.

She came to the interview with her beautiful purple summer *kimono* (Figure 8.4). Her silver short hair matched well with her *kimono* and sashes, by explaining her *kimono*:

> Most *ikebana* practitioners are well cultured. Through *ikebana* they know art and aesthetics, and they appreciate *kimono* like this. I know that they do not wear *kimono* as often as tea, but most of them are well cultured and elegant!

FIGURE 8.4 Women with *kimonos* for a tea gathering

Source: Photograph by the author.

She continued, sipping bitter Earl Grey tea:

> By the way, did you hear the news about Koizumi-*sensei*? She just received
> the cultural award with two other gentlemen this year. Her picture is all
> over the place in Akita. I heard that she will receive an award in the hotel
> next month. She became very famous in our town because of this. She was
> greeted by everyone in the town. She became famous, *erai* (greater) than
> her husband!

Kodama-*san* had been practicing *ikebana* and *chadō* for more than 30 years.
When I visited her place, she welcomed me in her *tatami* room. She had lost sub-
stantial weight after her husband's death, but she welcomed me with her smile.
She commented on a difference between *ikebana* and *chadō*:

> I think tea gatherings are more posh than *ikebana* gatherings. You see most
> of them [the attendees] wear *kimono*. They show their numerous utensils
> with certificates (*hakogaki*), and if you organize a private tea gathering, they
> even show their tea house and garden [Figure 8.5]. As you know, *ikebana*
> does not share *ikebana* container artist's names even if it is a famous one
> [Figure 8.6]. When I talk about *ikebana*, many of my friends comment that
> I am *suteki*, wonderful, and also they even say something like, I do these,
> so I am different from them. I do not feel I am different from them, but
> I do not deny it. I have a sense of pride that I am engaged with these trad-
> itional arts. Since I was little, let's say 60 years ago, my mother told me that
> women should do these things, and I am following what she said.

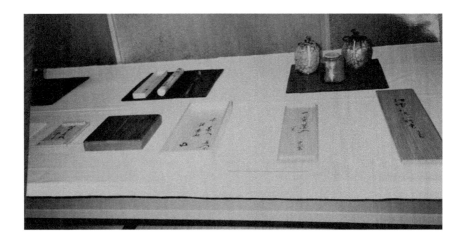

FIGURE 8.5 Utensils with certificates

Source: Photograph by Mihoko Chiba.

FIGURE 8.6 *Ikebana* exhibition

Source: Photograph by the author.

There are generally several *ikebana* exhibitions held in Akita City, especially in autumn. For example, 3rd November is recognized as the day for culture and registered as a public holiday. Every year, *ikebana* practitioners from different schools get together and hold exhibitions for about one week. Numerous practitioners come and visit other practitioners' work. The majority of guests seem to look more casual than those at *chadō* gatherings where many of them wear *kimono*.

Kodama-*san* talked about her relationship with her husband and *ikebana*:

> Yeah, he never complained about my *ikebana* lessons, rather he was very supportive. He commented that thanks to *ikebana*, we had so many flowers in our house. In fact, I heard that many family members support *ikebana*. First, they are very happy teaching from home, you know, they, like wives, are always at home.

As described briefly in the history chapter, I also had an opportunity to interview Chikuseikai *iemoto*, Terada-*sensei*, who became a female *iemoto* after her father:

> You see, my father took it for granted that I was going to be the next *iemoto*. He never told me that I was going to, but he told the public that I was going to be in the future. He never said a man should be the *iemoto*. Now I, as

the *iemoto* of Chikuseikai, attend numerous cultural meetings, which are still filled with men, and discuss the future of cultural promotion in Akita. I hope I am promoting the fruitful voices to Akita residents.

Status as madam, *MADAMU*

It can be interpreted that *ikebana* perpetuates social structures in class and gender among the senior generation in Japan. As my previous fieldwork examined, *chadō* is categorized as what Bourdieu (1984, 1987) calls high culture, defined as the taste of the dominant class (Chiba 2010). Much like *chadō*, this cultural form appears to be used as a tool for social distinction by the senior residents. The comments above maintain that practitioners tend to distinguish themselves as 'We as the people can devote or afford ourselves to flowers, not engaged with earning money.' However, it appears that this social distinction is not as apparent as in *chadō* for several reasons: 1. *ikebana* has offered *kyojō*, licensing and teaching qualifications more easily than *chadō*. 2. *ikebana* in general does cost in terms of flower materials, flower containers, exhibitions fees, and licensing, but not as much as *chadō*, which entails spending on numerous utensils, tea house interiors and gardens, *kimono*, and qualifications.

Ikebana teacher – gender empowerment

Ikebana can also be analyzed as a limited tool for gender empowerment. As mentioned above, *ikebana* modified its system to offer women the chance to become *ikebana* teachers or even *iemoto* in a flexible way. As a result, there are more female teachers and *iemoto* in *ikebana* than in any other traditional art field. However, *ikebana* is often taught at home, which does not threaten expected gender roles, men as *soto*, outside workers and women as inside housewives. Additionally, *ikebana* philosophy does not challenge the 'ideal' of women as the household manager by arranging flowers in their home. As often heard in the interviews, 'Her family did not mind her doing *ikebana*. They were all happy that they learned it since she can arrange flowers at her home.' On the other hand, Onme-*san*'s comment toward Koizumi-*sensei* suggests that *ikebana* provides a sense of gender empowerment to a certain extent. Through *ikebana*, Koizumi-*sensei* not only acquired cultural capital, but also status from her cultural award: symbolic capital. The interview suggested that practitioners assume that she acquired more respectable recognition than men, including her husband.

Iemoto – gender empowerment

Compared to *chadō* or other traditional arts, *ikebana* does provide a great sense of gender empowerment. It provides a woman an opportunity to be in a respectable position as an *iemoto*. This sense of freedom greatly empowers women within their home, and wider society. From my previous fieldwork in *chadō* society, I heard several voices from *chadō* practitioners saying that they felt a sense of the glass ceiling that women could not be the head of traditional arts . Influenced by the *ie* system in Japan, even among the artisan families, the first son or male is still preferred to be the head of the artisan world. As an example, only three among the ten artisan families directly working for the Urasenke tea school (*senke jyushoku*) have female heads (Tankosha 2021). This glass ceiling seems quite thin in the *ikebana* field. As described in the *iemoto* chapter, Kasumi and Toko became female *Iemoto* in Sogetsu school and Adachi school after their fathers.

Research questions

1. Westernization has brought certain gender roles to the world of *ikebana*. Do you know other gender roles that have changed because of westernization or modernization?
2. Should the traditions that involve certain gender stereotypes and gender roles change or should they be preserved as tradition? To what extent should they be accepted as tradition?
3. In such traditional art, even the gender-controversial part might be one of the parts which forms tradition. It seems to be difficult to change. How would we do it?
4. *Ikebana* appears to perpetuate ideas of gender expectations and social class. If *ikebana* practice became more affordable for people to participate in with a limited amount of money and time, would those ideas disappear?
5. To what extent should traditional art or culture be gender-neutral? (For example, *kabuki* does not allow females)
6. How does the concept of an ideal woman differ depending on social classes in Japan now?
7. What is the connection between gender disparity and gender issues and those traditional cultures in Japan? How about other countries? Are there any cultures that greatly shape their gender roles and characters?
8. Why does *ikebana* allow women to be *iemoto* while *chadō* does not?
9. Do you think young people see *ikebana* and tea gatherings as perpetuating social class/status?

Conclusion

This chapter argues that *ikebana* maintains gender roles and perpetuates class divisions to a certain extent. However, it should also be considered that *ikebana* provides a great sense of gender empowerment since opening the *iemoto* position to women. Academic discourse has researched the lives of factory workers, office ladies, and sex industry workers. However, Stalkers (2018) argues that *ikebana* and other traditional arts teachers have been missing from studies of women's labor in Japan. I argue that *ikebana*'s movement to open up the *iemoto* position for women is a significant shift in the intricate gender roles in Japanese society, and should be studied more, especially in Gender studies. Rappaport (1999: 256) argued that ritual is the reflection of the society in which it takes place, and that the ritual sometimes represents the opposite phenomena. *Ikebana* may represent the opposite to the phenomena of Japanese society in which the gender inequality movement and policy are significantly slow to change compared to other countries; in terms of gender gap, Japan was ranked 120th out of 156 countries in 2021 (World Economic Forum 2021).

References and further reading

Akitaken Kado Renmei (2007) *Akitaken Ikebanashi* [Akita Ikebana History], Akita: Akita Kyodo Press.

Agency for Cultural Affairs. (2018) *Heisei 29 Nendo Seikatsu Bunka Jittai Haaku Chosa Jigyo Hokokusho* [Heisei 29th Report of Fact-finding Survey Project About Life and Culture], Tokyo: Agency for Cultural Affairs.

Andressesn, C. and Gainer, P. (2002) 'The Japanese education system: Globalization and international education,' *Japanese Studies*, 22(2), 153–167.

Bourdieu, P. 1984 (1979) *Distinction: A Social Critique of the Judgment of Taste*, R, Nice (trans), Cambridge: Harvard University Press.

——— (1987) 'What makes a social class? On the theoretical and practical existence of groups,' *Berkeley Journal of Sociology*, 32, 1–17.

Fukuzawa, Y. (1899) 2001. *Onna-Daigaku*, Tokyo: Kodansha.

Goldsteir-Gidoni, O. (2005) 'Fashioning cultural identity: Body and dress,' in J. Robertson (eds.), *A Companion to the Anthropology of Japan*, UK: Blackwell Publishing Ltd, pp. 153–166.

Ito, Y. (2017) 'Meijiki iko no taishu ni okeru kado to jenda ni tsuite: joshi kyoiku no shiten kara' [Kado and gender in the mass since the Meiji Period: from the perspective of girls' education], *Waseda Studies in Social Sciences. Extra Issue, Students Journal in 2016 50-year Anniversary Edition for the Establishment of School of Social Science*, 187–196.

Jackson-Houlston, C. (2006) 'Queen lilies? The interpenetration of scientific, religious and gender discourses in Victorian representations of plants,' *Journal of Victorian Culture*, 11, 84–110.

Kato, E. (2004) *The Tea Ceremony and Women's Empowerment in Modern Japan: Bodies Representing the Past*, London: Routledge.

Kido, T. (2007) 'Sohonbuhokoku' [report from the general report], in Urasenke (ed.), *27kai Gakko Sado Tantosha Koshukai* [27th Tea Ceremony at School Seminar], Urasenke: Urasenke, pp. 2–17.

Kokubun, M. (2007) 'Dento, bunka wo sonchosuru kyoiku to gakuryoku mondai' [Education giving respect to traditional arts and its problem], in Urasenke (ed.), *Dai 27kai Gakko Sado Tantosha Koshukai* [27th Tea Ceremony at School Seminar], Urasenke: Urasenke, pp. 18–40.

Koyama, S. (2012) *Ryosai Kenbo: The Educational Ideal of 'Good Wife, Wise Mother' in Modern Japan*, The Netherlands: BRILL.

McLelland, M. J. (2012) 'Sex and censorship during the occupation of Japan,' *Asia Pacific Journal*, 10(37), 1–22.

McVeigh, B. J. (2005) 'Post-compulsory schooling and the legacy of imperialism,' in J. Robertson (eds), *A Companion to the Anthropology of Japan*, UK: Blackwell Publishing Ltd, pp. 153–166.

Nishiyama, M. (1962) *Gendai no Iemoto*, [Current Iemoto], Tokyo: Kobundo.

Rappaport, R. (1999) *Ritual and Religion in the Making of Humanity*, Cambridge: Cambridge University Press.

Stalker, N. (2018) 'Flower empowerment: Rethinking Japan's traditional arts as women's labor,' in Bullock J., Kano A., and Welker J. (eds.), *Rethinking Japanese Feminisms*, Honolulu: University of Hawai'i Press, pp. 103–118.

Sugimoto, Y. 2015 (1997) *An Introduction to Japanese Society*. 4th edn. Cambridge: Cambridge University Press.

Tankosha. (2021) 'Gendai no senke jyushoku' [Current tea artisans], *Nagomi*, 505, 5–38.

Terada, K. (2021) Personal communication 7 July.

World Economic Forum (2021). *Global Gender Gap Report 2021*, World Economic Forum.

9

TRADITIONAL ARTS EDUCATION

From the previous chapter, we explored that the majority of *ikebana* practitioners are middle-class senior women who may have started from their bridal training. We have discussed that there are few from the younger generations who are engaging with traditional arts, including *ikebana*. This chapter describes how *ikebana* and other Japanese traditional arts have mainly been taught in Japanese compulsory education. As described in the previous chapter, *ikebana* education is not offered as a regular subject in Japan. It tends to be taught dedicated to non-subjects, particularly integrated studies (*sōgō gakushū*), moral education classes (*dōtoku*), and special activities (*tokubetsu katsudō*); as curricular club activities (*kurabu katsudō*) or extracurricular club activities (*bukatsudō*). This chapter analyzes how this traditional art education differs according to varying teaching styles and regions in Japan. In order to explore effective teaching styles, the chapter shares interviews from teachers and those who engage with traditional arts education. Along with direct insights given by teachers, staff, and students, the chapter will conclude by introducing the potentials of traditional arts/*ikebana* education in Japanese compulsory education.

Current trends of traditional arts education

Through the latter half of the twentieth century, the rapid decline of interest towards traditional arts became ever more concerning (Kokubun 2007). Considering these worries, Kokubun (2007) argues that the Japanese government decided to add the statement in Article 2 in the Fundamental Law on Education (*kyouiku kihon ho*); Japanese education aims to respect its tradition and culture, respect other nations, and promote peace and development in the globalized community. Thereafter, the Ministry of Education, Culture, Sports,

DOI: 10.4324/9781003248682-10

Science, and Technology (MEXT) started to increase traditional arts education, together with other educational reforms, from 2006 onwards (Kokubun 2007, Hashimoto 2015, Yoshida 2016). The same statement was inserted (as section 21(3)) into the School Education Law in 2007. The 2010 Curriculum Guidance requests the emphasis of education related to tradition and culture to promote the appropriate Japanese person in globalized society (Yoshida 2016). Curriculum Guidance in 2011 describes traditional arts–related education in detail: education should emphasize Japanese instruments in music class, traditional lifestyle and culture in home economics, and Japanese painting in art class (Yoshida 2016). Followed by the statement of 120-2-5 law, this guidance emphasizes that every local board of education should create their own education plan considering the local conditions (Yoshida 2016).

Kokubun (2007) also states the reform was relevant to the government policy in the global context: Japanese traditional arts promotion for increasing foreign tourists. Additionally, the government emphasized the significance of Japanese identity to interact with global business and trade (Kokubun 2007, Hashimoto 2015). Due to this reform in 2006, traditional music including Japanese harp, guitar, and drum (*taiko*) were to be encouraged and taught more in music classes in compulsory education as only western music had been taught since the Meiji period (Yoshida 2016). Moreover, MEXT also made Japanese martial arts a compulsory subject (for both genders) for junior high and high school students from 2012 (MEXT 2011).

Due to these reforms, traditional arts classes have gradually offered it as part of integrated studies, special activities, as a club activity, and in moral education classes (Kido 2016). Integrated studies were promoted under the *Yutori kyōiku* education philosophy, which, from 2000, MEXT encourages a form of relaxed education with 'room to breathe.' Curricular club activities are held regularly, once a week for 45 minutes from 4th grade in elementary school. Students can select their favorite club activities offered by their school. Moral education class is a compulsory subject for every year group. It is offered once per week for 45 minutes at elementary school. Students are encouraged to learn respect towards rules, justice, family, groups within society, local communities, the nation, international groups, and traditional arts (Yoshida 2016). There are numerous extracurricular club activities in elementary and junior high schools in Japan and it is common for junior high school students to attend. Japanese students are well known to engage with these activities after class or at the weekend. There are several sports and art clubs: baseball, soccer, tennis, basketball, *judō kendō*, art, science, and music. Some of the schools also include *ikebana* clubs as an option (Hendry 2019).

According to Ando, the head of education department in Ikenobo (2021), the number of schools teaching *ikebana* has not changed much at the junior and high school level, but this number is increasing in the elementary school and preschool level. Ando (2021) states that about 1850 educational institutions are engaged with Ikenobo *ikebana*. The revised curriculum guidance from 2017 is followed, this guidance newly emphasizes education related to traditional arts along with computer programming and English education (MEXT 2017).

Kyoto

The number of the traditional arts club activities and classes varies depending on the region in Japan. Kyoto, recognized as the center of the traditional arts, emphasizes teaching traditional arts and offers more traditional art classes than other prefectures. Kyoto city constructed a strong volunteer system using local communities to provide traditional arts instructors and supporters. With these support systems, the city provides traditional arts classes (Kyoto City Board of Education 2020).

Twenty percent of National treasures, 15 percent of cultural heritage sites, and numerous headquarters of temples, shrines, and traditional arts are based in Kyoto City. Considering these facts, the Agency of Cultural Affairs relocated from Tokyo to Kyoto in 2021. Because of this relocation, the Kyoto City Board of Education decided to offer *chadō* classes to all elementary school students and *ikebana* classes for all junior high school students in Kyoto City in 2019 (Kyouiku Shimbun 2019). This will be a pilot project until 2021, after which Kyoto City aims to implement them, and these traditional arts classes are likely to become mandatory subjects in compulsory education from 2022. The Kyoto City Board of Education has a budget of approximately 10,000,000 yen for equipment and hiring instructors. These traditional arts are taught as part of the moral education classes (Kyoto City Board of Education 2020). How do other places engage with traditional art education?

Akita

Akita City, located in the Tohoku area, is the capital city of Akita Prefecture. Although this place is well known as a rice-producing farming community, these days, 80 percent of the residents are involved in tertiary industries, which include wholesale and retail trades and services (Akita Prefectural Government Statics Division 2015). Akita Prefecture compulsory education has been ranked as one of the highest in Japan in results on standardized tests (Akita Prefectural Government, Compulsory Education Division 2020). Akita City offers *ikebana* classes as part of integrated studies, curricular or extracurricular club activities, but the numbers of the activities are limited. There are 41 elementary and 23 junior high schools in Akita City, however, only two elementary schools taught *ikebana* in 2020 (Terada 2021).

According to Sato-*san* (2020), the director at the Akita City Board of Education, Akita City tends to focus on local festivals, farming culture, traditional art works, and historical events in integrated studies. For instance, Hiroomote elementary school offers local festival studies as part of integrated studies for third grade students (Figure 9.1). During the first couple of classes,

FIGURE 9.1 Local festival class in Akita

Source: Photograph by the author.

students first learn from local residents about the history and meaning of a particular festival. They also attend the festivals and carry a float to the shrine with
support from the local community. This festival was historically open only to
males, but is now also open to female students (Sato 2020). On the other hand,
Tsuchizaki elementary school offers local history classes in integrated studies.
As the Tsuchizaki area is well known as the last place to be attacked during the
Second World War, the local students learn more about their local history by
visiting a museum and interviewing the residents who experienced the attack,
and then they prepare presentations (Sato 2020). While there is no junior high
school offering *ikebana* courses in Akita City, there are two junior high schools
offering classes in *chadō*, both of which are the combined junior and senior high
school style. Akita Minami Combined High School offers *chadō* education as
their extra-curricular activities, and Goshono Gakuin School offers *chadō* classes
as their integrated studies (Wakamatsu 2020.

Nishin School teaches *ikebana* as part of the curricular club activity. According
to Terada-*sensei* (2021), one of her students, Kondo-*san*, applied for a project to
promote teaching traditional arts supported by the Cultural Affairs. Her application was accepted, so she asked her local elementary school if she could offer
ikebana classes, and the school accepted the offer. Kondo-*san* commented that this
Cultural Affairs project was the start of having *ikebana* classes as one of the club

activities. Even after the financial support from Cultural Affairs ended, Nishin School and Kondo-*san* were willing to support the *ikebana* classes and have still continued to teach them for ten years now. Ten sessions are offered every year, and *ikebana* classes had 15 students in 2021 –three of them boys. *Ikebana* students from Nishin School offer *ikebana* presentations once a year at the Chikuseikai *ikebana* exhibition (Figure 9.2). Parents and friends of these students visit the exhibition to support the children's *ikebana* work. Terada-*sensei* commented that this presentation is a significant opportunity to attract different generations to *ikebana*.

The question might arise as to how the case of traditional arts education in Akita city is relevant elsewhere in Japan. While the reference to the

FIGURE 9.2 *Ikebana* presentation from Nishin School

Source: Photograph by the author.

historical, cultural, and private school background may not resonate with those in metropolitan areas, Terada-*sensei* and Sato comment that it may be possible to identify similar traditional arts education cases in Akita with those in other non-metropolitan areas other than Kyoto as the center of Traditional arts and festivals, and Okinawa, which retains a distinctive history and culture (Hendry 2019).

Combined junior and high school style

Sato-*san* (2020) and other interviewees state that teaching traditional arts is more accessible in combined junior and high school than junior high school. For example, Hakuo Public School in Tokyo changed to a six-year curriculum from 2004. This school has offered the traditional arts including *chadō* and *ikebana* as selective subjects (Valxl 2020). Goshono Gakuin in Akita City was opened as a combined junior and high school in 2000. It offers local studies (*kyodo gaku*), biweekly 100-minute classes, for which all students select a topic from 13 choices, including Chinese, computing, pottery, Japanese harp, and *chadō*. Kawaguchi-*sensei*, a retired teacher as well as a *chadō* teacher, teaches *chadō* for about 20 students in the school's *tatami* room. Local studies are offered to all students except the 3rd year seniors. Sato-*san* (2020) states that this difference is due to the high school entrance exam; junior high schools in Japan generally start to prepare entrance exams from the autumn semester in the second grade. Combined junior and high schools' curriculums are more flexible and have more time to offer traditional arts education.

Private girls school

Ninety-nine percent of elementary schools and 93 percent of junior schools are public schools in Japan (Valxl 2020), and most of the private schools are based in the Kansai and Tokyo metropolitan areas. According to the survey from Valxl, 34 percent and 30 percent of private schools in the Kansai and Tokyo metropolitan areas, respectively, offered *ikebana* clubs in 2020. For public schools, the proportion is 8 percent and 0 percent. This data indicates that traditional arts education is more popular among private than public schools. Kadowaki-*sensei* is a teacher working at an elementary school in Akita City, **and** she shared her thoughts:

> Well, you see I think the private schools have the atmosphere to engage with these art forms. No parents complain about that, rather they support them since they have done that. In fact, I heard that some alumni members support their studies by donation and volunteer support.

Kadowaki-*sensei* (2020) suggests that this disparity indicates that *ikebana* is still perceived as a status enhancing pursuit, in relation to the social class issues in Japan.

In terms of gender, 0 percent and 3 percent of the boys schools and 36 percent and 54 percent of the girls schools in Kansai and in Tokyo metropolitan areas, respectively, were offered an *ikebana* club in 2020 (Valxl 2020). This data seems to be related to the fact that *ikebana* continues to be associated with girls and young women. It is significant that *ikebana* clubs or classes are attended by more girls than boys in mixed schools. Terada-*sensei* stated:

> Oh, yes, there are hardly any boys, this is strange, we never ever said that this is for girls. But every year, all or most of the students are girls. If we see boys, we say, 'Oohhhh, *mezurashi!* (unusual).'

Terada-*sensei* further comments that all of the members in 2020 in Akita Kita High School and Yuzawa Shohoku High School were girls, and she further states that most of them are girls every year.

This section has described how the traditional arts have been taught in Japanese compulsory education. These art forms tend to be taught mainly as special classes, aside from compulsory subjects. Additionally, the number of traditional arts classes varies depending on the region in Japan. According to Sato-*san* (2020) and Wakamatsu-*sensei* (2000), to some extent, educational content and policy are decided by the local boards of education. The approach is also different depending on the type of junior high school. Then, what kind of other future agenda do we have for promoting traditional art education in Japan?

Future agenda

Instructors: The Agency of Cultural Affairs (2018) states that the lack of instructors is the issue. Shiroma (2013) analyzes the possibility of teaching Japanese traditional performing arts in Japan, and highlights this issue. For this matter, Kyoto City organizes and uses the volunteer system. Around 8,000 local residents who can teach *ikebana*, *chadō*, *noh*, *kabuki*, art works, and festivals are registered, and they offer to teach different art forms at schools (Kyoto City Board of Education 2020). On the other hand, the NPO (Non-Profit Organization) called *Musubinokai* offers similar support. Founded in 2010, this NPO aims to promote traditional art education at schools and is organized with researchers, teachers, and local residents who can teach traditional arts (Morita 2020). Yoshida (2016) comments that the current compulsory education policy promotes the school within the local communities. It appears that receiving support such as extra teachers from the local communities seems to be the ideal model for promoting traditional arts education. Sekine (2016: 99) states that there is a need for supporting traditional arts education. He criticizes that there is not enough financial support given to each traditional art institution that supports this policy which advocates for community engagement in Japan.

In terms of these issues, Terada-*sensei* and Kondo-*san* commented that *ikebana* instructors should be paid for their teaching concerning career opportunities for the younger generation. Terada-*sensei* shared:

> I think this matter should be considered seriously. We have been teaching for free all our lives. But if we carry on, younger generations will think there is no money in this business and avoid becoming *ikebana* teachers.

By listening to their life stories, it appears that schools tend to take it for granted that their work is voluntary.

Parents' understanding toward traditional arts: A survey was conducted in Kyoto City to investigate parents' points of view, inquiring about their ideal weekend activities for their children. About 2,700 elementary school parents and 700 junior high school parents answered, and out of 23 selected activities, including sports and outdoor activities, traditional art activities including *chadō* and *ikebana* ranked 21st in 2016. Tooyama (2016) stated that even in Kyoto City, which has a strong connection with traditional arts, parents have less interest in these art forms. Tooyama (2016) comments that without these preferences, schools should engage with the subject which they believe is appropriate for students.

Compared with other traditional arts: Calligraphy (*shodō*) is practiced as a compulsory subject and the extension of *kokugo* in elementary and junior high school (Beomjin 2012) (Figure 9.3). Generally, this class starts from the 3rd grade and occurs once a week for 45 minutes in elementary school. Calligraphy classes were also not encouraged after the Second World War because of the feudalistic sentiment, only becoming available from 1971 onward. Kobayashi (2003) introduced an experimental *kabuki* course in high school in Japan. However, other Japanese traditional performing arts includes *noh*, traditional Japanese dance are still not commonly taught as a subject or club activities in compulsory education. On the other hand, *chadō* classes are offered more than *ikebana* classes. As mentioned above, while there were only two schools teaching *ikebana*, seven schools were offering *chadō* education in Akita City in 2020 (Urasenke Akita 2020).

Cultural nationalism?

Applying Befu's (2001) characterization of cultural nationalism as focusing on a national identity shaped by cultural traditions and by language, it appears that this education is related to cultural nationalism. Traditional arts education is related to identity. As stated above, the course of study from 2010 requires traditional arts education to promote the ideal *Japanese person* in the globalized society. Sumioka (2016) states that traditional arts education is indeed relevant in

FIGURE 9.3 Calligraphy written by 5th grade students

Source: Photograph by the author.

constructing a Japanese student's identity. To explore these identities, the concept of way, *do* philosophy derived from Taoism and Buddhism in *ikebana* is also highlighted to enhance the uniqueness of Japanese people. In order to establish Japanese national identity, the Meiji government attempted to utilize arts and language to unite national identity (Surak 2012). It appears that similar tendencies are seen in this Reiwa period.

The question arises whether this promotion of traditional arts education is relevant to Nikkyoso, and Nippon kaigi. The Nikkyoso is the teacher's union founded post war, which tends to believe in left wing perspectives: it is against raising the Japanese flag and singing the anthem. This tendency comes from the strong will never to send their own students to war as soldiers. The Nippon kaigi was established in 1998, and works to promote culturally conservative right-wing philosophy (Fujiu 2017). Former Prime minister Abe is the chair of this Nippon kaigi. According to Fujiu (2017), Nippon kaigi has six major goals, and one of them is to promote Japanese identity through education. Their promotion of tradition dovetails with the reinforcement of traditional gender roles (Fujiu 2017) to an extent which is seen as discriminatory to contemporary sensibilities. Nonetheless, they were a visible and effective vehicle promoting the 2006 Fundamental Law of Education, which emphasized respect for the nation. Abe also used the idea of tradition to castigate alternative educational reforms proposed by the Nikkyoso. Nikkyoso gathered 8 million signatures to oppose the education reform (Fujiu 2017).

Reproduction of class and gender

Similar to *ikebana* practitioners' background, *ikebana* and other traditional arts education can also be analyzed as perpetuating class and gender stratification. The statistics described above, dominated by private and elite combined schools and by girls' schools, along with the comments from interviews seem to be relevant to Bourdieu's argument that cultural form is used as a social distinction tool. Kadowaki (2020) comments that parents of private school children have more understanding towards culture, such as traditional arts, which does not connect directly to their children's economic capital. Bourdieu (1984, 1987) also comments that middle-class people tend to engage with tastes which do not directly improve their financial position. In regard to perpetuating gender roles, the term 'bridal training' is dying out. However, Japan is still recognized as a society which continues to hold clear gender assigned roles and concepts of ideal masculinity and femininity (Liddle 2000, Chiba 2010, Ueno 2020). As can also be perceived from the term *jyoshi ryoku*, ideal Japanese femininity is related to a girl or young lady who has substantial etiquette and manners, skills for cooking, sewing, cleaning, and other domestic work (Roberts 2005, Ueno 2021). This ideal femininity seems to remain, however, as a catchphrase for *ikebana* education in private girls' school; *jyosei no hinkaku wo sodatemasu* (improve appropriate women's dignity). Okano and Tsuchiya (1999) analyzed how gender, poverty, and elitism influence education in Japanese society. Indeed, traditional art education, especially *ikebana*, is heavily related to gender and elitism. Additionally, traditional art education phenomena clearly highlight the fact that Japanese education still follows the deviation score: tool to measure the deviation of scores on tests from the average score (*hensachi*) (Goodman and Oka 2018). As has been commented above, six-year combined junior high and high schools tend to have more traditional art education than regular three-year junior high schools (elementary and junior high school are the compulsory education in Japan and most Japanese students attend three-year junior high school and three-year high school). Sato (2020) comments that the junior high school teachers always complain they just don't have enough time for preparing their students for the entrance exam. Teachers are focused on making sure each student understands the material before moving on to the next point, but are also driven by a deadline, the exam dates: teachers are often forced to teach to the exam rather than to the students.

Research questions

1. What is the advantage of teaching traditional arts for all students regardless of gender?
2. What is the difference in attitude toward traditional arts between students of private girls' schools and of public schools? Is there any difference in the ways they perceive the roles of those arts?
3. Will Japanese traditional arts, which are often perceived as the aspects of a country's identity or culture only available to the rich, still remain as a culture or just a commodity for the rich?
4. Even though the notion of bridal training has faded out from Japanese society, there are still clear gender roles. What kind of roles women have today do you think are the legacy of the notion of bridal training?
5. If *ikebana* became mandatory in compulsory education, would it make any difference to the people's image of *ikebana*? Would that be good for future generations?
6. How does the current traditional arts education shape children's identity as Japanese and as women/men?
7. *Ikebana* and *chadō* are historically considered to be for women. Do you think they should be taught in school? Why?
8. How can we teach traditional art education without perpetuating class and gender roles?
9. For children, what is the meaning of learning traditional arts, including *ikebana*?
10. Will this traditional art education serve to preserve the traditional arts?

Conclusion

This chapter focused on current issues surrounding traditional arts, especially *ikebana* education in compulsory education, drawing on the voices of people who engage directly with this education. Numerous issues related to traditional arts education including the contents of integrated studies and moral education are not officially recorded or documented in the local boards of education. So, direct voices from teachers, staff, and students were crucial for this study.

Traditional arts education is not offered as a regular subject in Japan. It tends to be taught in integrated studies, moral education, special activities classes, club activities, and extracurricular activities. Yet, this *ikebana* education is increasing after the educational reform in 2006. Considering the example of Kyoto City, it appears that this education has increased to a significant extent. However, the approach toward traditional arts education varies greatly depending on the region and style of the school. It is practiced more at combined junior high and high schools, private schools, and girls' schools in Japan. For this reason, it can be argued that this traditional art education still serves to perpetuate class and gender roles, and may additionally be connected to cultural nationalism.

Nonetheless, despite the concerns, there are few who wish to see *ikebana* disappear. It is interesting to consider how traditional art education will be recognized in the future. With the globalized context, it appears that the number of classes will increase. Current *ikebana* education is supported by local senior residents who learned it in relation to their bridal training. Yet, the training of qualified teachers for the next generation in traditional arts will be a significant agenda item to consider in the future.

References and further reading

Agency for Cultural Affairs. (2018) *Heisei 29 Nendo Seikatsu Bunka Jiitai Haaku Chosa Jigyo Hokokusho* [Heisei 29th Report of Fact-finding Survey Project About Life and Culture], Tokyo: Agency for Cultural Affairs.

Akita Prefectural Government Statics Division. (2015) *Akita Statics in Brief 2015*, Akita: Akita Prefectural Government.

Akita Prefectural Government, Compulsory Education Division. (2020) Personal Communication 10 December.

Ando, T. (2021) Personal Communication 13 July.

Befu, H. (2001) *Hegemony of Homogeneity: An Anthropological Analysis of Nihonjinron*, Melbourne: Trans Pacific Press.

Beomjin, C. (2012) 'Calligraphy education with a writing brush in the Courses of Study of South Korea and Japan,' *Art Education Association*, 33, 301–314.

Bourdieu, P. 1984 (1979) *Distinction: A Social Critique of the Judgement of Taste,* R, Nice (trans). Cambridge: Harvard University Press.

——— (1987) 'What Makes a Social Class? On the Theoretical and Practical Existence of Groups,' *Berkeley Journal of Sociology*, 32 , 1–17.

Chiba, M. (2020) Personal Communication 1 December.

Chiba, K. (2010) *Japanese Women, Class and the Tea Ceremony*, London: Routledge.

Fujiu, A. (2017) *Document Nihon Kaigi*, Tokyo: Chikuma Shinsho.

Goodman, R. and Oka, C. (2018) 'The invention, gaming, and persistence of the hensachi standardised rank score in Japanese education,' *Oxford Review of Education*, 44(5), 581–598.

Hashimoto, T. (2015) 'Dento bunka kyoiku ni kakawaru zoukeikatsudo no kakuchoteki gakushu toshiteno kanosei [Research on the possibilities of expansive learning of the artistic activities for traditional culture education], *Journal of Hokkaido Kyoiku University*, 66(1), 253–268.

Hendry, J. (2019) *Understanding Japanese Society*, 5th edn. London: Routledge.

Hirota, T. (2004) *Kyoiku Gensetsu no Rekishi Shakaigaku* [Historical Sociology in Education], Nagoya: Nagoya University Press.

Hsu, F. (1963) *Clan, Caste and Club*, Princeton, New Jersey: Van Nostrand.

——— (1975) *Iemoto: The Heart of Japan*, New York: New York: Halstead.

Kadowaki, S. (2020) Personal Communication 15 November.

Kawaguchi, K. (2020) Personal Communication 13 December.

Kido, T. (2007) 'Sohonbuhokoku' [report from the general headquarters], in Urasenke (ed.), *Dai 27kai Gakko sado tantosha koshukai* [27th Tea Ceremony at School Seminar], Urasenke: Urasenke, pp. 2–17.

——— (2016) 'Shusaisha aisatsu' [Greeting from the organizer], in Urasenke (ed.), *Dai 36kai Gakko sado tantosha koshukai* [27th Tea Ceremony at School Seminar], Urasenke: Urasenke, pp. 8–13.

Kobayashi, Y. (2003) 'Nihon no dentougeinou wo gakkoukyouiku ni dounyuu suru kanousei to kadai Ashikaga Minami Koutougakkou sougougakusyuu "Kabuki Kouza" wo Jirei to shite' [Possibility of teaching Japanese traditional performing arts in school curriculum], *Research Journal of JAPEW*, 20, 55–65. DOI: 10.11206/japew2003.2003.55

Kokubun, M. (2007) 'Dento, bunka wo sonchosuru kyoiku to gakuryoku mondai' [Education giving respect to traditional arts and its problem],' in Urasenke (ed.), *Dai 27kai Gakko sado tantosha koshukai* [27th Tea Ceremony at School Seminar], Urasenke: Urasenke, pp. 18–40.

Kyoto City Board of Education. (2020) Personal communication 27 November.

Kyouiku Shibun. (2019) 'Shiritsusho, chude sado, kado no taiken jyugyou, zenko jichishie' [Tea ceremony and flower arrangement at municipal elementary and junior high schools. Will be implemented in all schools in Kyoto city] www.kyobun.co.jp/news/20190213_03/ (accessed 10 November 2021).

Liddle, J. (2000) *Rising Suns, Rising Daughters: Gender, Class and Power in Japan*, London: Zed.

Ministry of Education, Culture, Sports, Science, and Technology. (2011) *Revised School Curriculum.* www.mext.go.jp/a_menu/shotou/new-cs/youryou/1282000.htm. (accessed 10 November 2021).

Ministry of Education, Culture, Sports, Science, and Technology. (2017) *Revised school curriculum.* www.mext.go.jp/content/1413522_002.pdf. (accessed 10 November 2021).

Morita, Y. (2020) *Yokoso Tentogeino no Sekai* [Welcome to Traditional Art World], Tokyo: Kunpusha.

Okano, K. and Tsuchiya, M. (1999) *Education in Contemporary Japan: Inequality and Diversity. Contemporary Japanese Society*, Cambridge: Cambridge University Press.

Roberts, G. (2005) 'Shifting contours of class and status,' in J. Robertson. (ed.), *A Companion to the Anthropology of Japan*, London: Blackwell Publishing, pp. 104–123.

Sato, T. (2020) Personal Communication 10 November.

Sekine, T. (2007) 'Sohonbuhokoku' [report from the general report], in Urasenke (ed.), *Dai 27kai Gakko sado tantosha koshukai [27th Tea Ceremony at School Seminar]*, Urasenke: Urasenke, pp. 78–89.

——— (2016) 'Sohonbuhokoku' [report from the general report], in Urasenke (ed.), *Dai 36kai Gakko sado tantosha koshukai [36th Tea Ceremony at School Seminar]*, Urasenke: Urasenke, pp. 98–100.

Shiroma, S. (2013) 'Senmonka tono renkeiniyoru dento, bunkano kyoikuni kansuru kyouin no katari' [The teachers' talk in terms of traditions and cultures by cooperation with experts], *Japanese Association of Education Psychology*, 55:5.

Sumioka, H. (2016) 'Education by means of Japanese traditional performing arts—Toward fostering the Japanese fundamental culture,' *Journal of Osaka Aoyama University*, 8, 25–37.

Surak, K. (2012) *Making Tea, Making Japan: Cultural Nationalism in Practice*, Stanford: Stanford University Press.

Terada, K. (2021) Personal Communication 25 July.

Tooyama, H. (2016) 'Kyoto furitsu koko deno sado gakushuu no torikumi nit suite' [Tea ceremony education in Kyoto prefectural high school],' in Urasenke (ed.), *Dai 36 kai Gakko sado tantosha koshukai* [36th Tea Ceremony at School Seminar], Urasenke: Urasenke, pp. 58–75.

Ueno, C. (2020) *Ueno-sensei, Feminism ni tsuite zero kara oshietekudasai* [Please teach me Feminism], Tokyo: Yamato Shobo

——— (2021) *Ofukushokan, Genkaikara Hajimaru* [It starts from the bottom], Tokyo: Gentosha.

Urasenke Akita Branch. (2020) Personal Communication 10 November.

Valxl (2020) *Sadoubu ga aru chugaku (The Junior High School which has Tea Ceremony Club)* www.study1.jp/kanto/special/club/cultural/club.html?c=tea (accessed 10 November 2021)

Wakamatsu (2020) Personal Communication 13 December.

Yoshida, K. (2016) 'Gakko shidosha no yakuwari [role of tea ceremony teacher at school],' in Urasenke (ed.), *Dai 36 kai Gakko sado tantosha koshukai* [36th Tea Ceremony at School Seminar], Urasenke: Urasenke, pp. 32–57.

Yoshimura, M. (2013) 'English language activities as part of community-based world heritage learning,' *Nara University of Education Journal*, 22, 217–222.

Yamamoto, T. (2020) Personal Communication 14 December.

Zhan, G. (2019) 'Gurobarukashakai ni okeru nihon no dentoubunka no jittai' [The Reality of Traditional Japanese Culture in a Globalized Society], *Research Institute for Studies in Arts and Sciences*, 5, 63–81.

10

UTENSILS AND *IKEBANA* ARRANGEMENTS

There are numerous styles of *ikebana* arrangement depending on the occasions and seasons. In *ikebana*, practitioners not only emphasize the types of flower, but also the flower containers (*kaki*). This chapter will closely focus on *ikebana* utensils and basic techniques. By exploring how to cut and preserve flowers and branches, readers will further learn some typical *ikebana* arrangements in the style of Ikebono. The rest of the chapter shares some guidance on how to make *ikebana* courses fruitful and productive: introducing tips such as possible field trips and online courses.

How to prepare and preserve flowers

Every flower and branch is different, when *ikebana* is practiced, extra materials also tend to be prepared. Generally, teachers source the materials from a specific flower shop or their gardens, fields, or nearby mountains. For beginners, some branches and flower combinations are encouraged, as the strength of the branches complement the delicate beauty of flowers well and offer well-balanced elegance (Takenaka 1995). When the plants are gathered, it is appropriate to place them in deep water as soon as possible and store them in a cool room, avoiding direct sunlight. Fresh tap water should be used for *ikebana* to avoid any bacteria. If the flowers have wilted, let them soak in a bucket with slow-running tap water. Before placing anything in the water, it is also recommended to cut the stem diagonally to increase the surface area and expand water intake. There are other techniques to stimulate water intake: mint water for a clematis (*tesen*), combining water with a small amount of alcohol, *sake* or whiskey for wisteria (*fuji*), burnt alum for bush clovers and hydrangea, vinegar for pampas grass (*susuki*), and using a syringe to inject water for lotuses (*hasu*) (Okada 1992, Kawase 2004).

DOI: 10.4324/9781003248682-11

FIGURE 10.1 How to cut

Source: Photograph by the author.

What to prepare

Scissors (*hasami*): There are several different kinds of scissors (Figure 10.1). The figure below shows the most commonly used. They can be purchased for around 2,000 yen. Unlike *chadō*, where the tea procedure is strictly right-handed, flowers can be cut with the left hand in *ikebana*. If it is necessary, left-handed scissors for *ikebana* are also available. A small saw is also useful for cutting thick branches, especially for *rikka* style (Takenaka 1995).

Kenzan: It is a spiked holder in which flowers and branches are placed (Figure 10.2). There are many sharp narrow spikes on the *kenzan*, so practitioners have to be careful not to stab, stick, or prick their fingers. These *kenzan* can be purchased for around 1,000–2,000 yen. A *kenzan* is recommended for beginners, however, other metal materials such as simple branches can be used as alternatives as they practice. There are a variety of *kenzan*, with small, big, round, and square shapes available. The *kenzan* should not be apparent in the final arrangements, so practitioners tend to hide them with leaves or stones (Sparnon 1991).

> **Wires:** Wires are used to support, straighten, curve, or bend the branches and flowers (Figure 10.3). These are well used for *rikka* arrangements.
> **Paper tape:** Tape is used to amend the branches or hide the wires along the branches, and the color should match the materials.
> **Flower container, *kaki*:** A generic flower container is called *kabin*, but it is called *kaki* in ikebana. *Ikebana* practitioners pay attention to finding the perfect balance between the *kaki* and the flowers. It is believed the beauty of the flowers is enhanced by the *kaki*. There is a variety of *kaki* in *ikebana*. Here are some different shapes:
> **Shallow basin-shaped (*suibangata*):** There are round, oval, rectangular, or triangular shapes. This *kaki* is used for beginners (Figure 10.4).

FIGURE 10.2 *Kenzan*

Source: Photograph by the author.

FIGURE 10.3 Tape, wire, and scissors

Source: Photograph by the author.

FIGURE 10.4 *Suibangata*

Source: Photograph by the author.

FIGURE 10.5 *Compotgata*

Source: Photograph by the author.

FIGURE 10.6 *Tsubogata*

Source: Photograph by the author.

FIGURE 10.7 *Tsutsugata*

Source: Photograph by the author.

Compote-shaped (*compotogata*), bowl-shaped (*hachigata*): This container has a stand, and is suitable for the tall-styled *ikebana* arrangements (Figure 10.5).

Jar-shaped *(tsubogata)*, cylinder-shaped *(tsutsugata)*: For these containers, practitioners do not use a *kenzan*. Instead, they use other techniques to support flowers and prevent them from moving around in the container (Figures 10.6 and 10.7).

Bucket and towel: It is important to prepare a decent-sized bucket with a substantial amount of water. Flowers and branches should be cut in the water to provide sufficient water intake. Towels are used to clean up the area where this is practiced.

Water sprayer: It is useful to have a water sprayer for applying moisture to leaves, branches, and flowers, especially for *ikebana* exhibitions, which take place in air-conditioned rooms. Spraying can also get rid of some dirt.

Different kinds of formality

Formality shin, gyō, sō

There are three different levels of formality: *shin, gyō, and sō*, meaning formal, semi-formal, and informal respectively. Bronze and porcelain flower containers are recognized as the most formal (*shin*) utensils. Glazed pottery made in Japan, such as Seto, Tanba, and Takatori, is defined as semi-formal (*gyō*). Bamboo, wicker baskets, and unglazed Japanese pottery including Bizen, Shigaraki lacquerware, and glass flower containers are considered as the informal flower container (*sō*) (Kawase 2000). These three different formalities were originally derived from calligraphy, and are also used to indicate the formality of bowing, *kimono* wear, and utensils used in *chadō*. However, it does not mean that the *shin* arrangement should be most respected, as even the most informal style expresses the refined flower arts.

Glass and basket containers can be used for expressing coolness. There are different qualities and shapes of baskets. The refined Chinese basket or the Japanese reproduction of these is considered to be the formal form of *sō*. Lacquerware surfaces, especially antique ones, are considerably sensitive, so water should be avoided and only a soft cloth dampened with warm water should be used for cleaning. Glass or plastic water containers (*otoshi*) are generally used as the inner water vessels of the lacquerware. There are various styles of bamboo containers: single cut, double cut, flute-shaped style, and boat-shaped style. Except for the fresh blue bamboo container, the rest use an *otoshi*. A porcelain container is delicate, especially with cold water, so it is recommended not to leave the water in this container during the winter season (Lim 2009). Map 10.1 below shows the locations of different potteries within Japan.

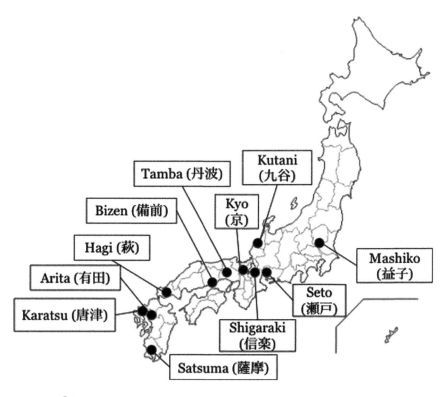

MAP 10.1 Japanese pottery map

Source: Kawase 2000.

Way to approach uniqueness

Through *ikebana* containers, practitioners express their wishes and creativity. For example, wishing for peace in the world, the next grandmaster of *Ikenobo* arranged *ikebana* with hand grenades (Ikenobo Kadokai 2013). Due to the lack of iron in Japan at the end of the Second World War, hand grenades were made from Shimizu, Shigaraki, and Bizen ware (Ikenobo 2013) (Figure 10.8). Another practitioner used a shoe to express a new step toward his future life. For *ikebana* freestyle, any container is acceptable in expressing one's creativity (Figure 10.9).

How to cut flowers

Hold the upper handle with the thumb and lower handle with the rest of the hand, and cut the materials by clenching hands. Branches should be cut diagonally, so that the edge can be easily inserted into the *kenzan*. Flowers and leaves should be cut horizontally (Fujiwara 1976).

FIGURE 10.8 Hand grenade flower arrangement. Copyright Ikenobo

FIGURE 10.9 Shoe arrangement

Source: Photograph by the author.

How to bend flowers

Folding (*oridame*): These techniques are used mainly for branches for *rikka* style (see more details in the History Chapter) (Figure 10.10). The branch is slit at the point you want it to bend and then folded slowly. To support the angle, tape and wires can be used.

Squeezing (*nigiridame*): A squeezing technique is taught in the basic *ikebana* training (Figure 10.11). Simply use your fingers to squeeze both sides of the area you wish to bend.

Stroking (*shigokidame*): This technique is suitable for long slender leaves. Hold the leaf between the thumb and index finger and stroke it from the area you wish to the leaf tip.

How to arrange *ikebana*

There are a variety of *ikebana* arrangements. Six different arrangements are introduced below. For most occasions, the front part of the flowers, where the sun is facing, should be checked to ensure they face the front part of the *ikebana* arrangement (Kawase 2000). Look carefully at the angle of branches and decide which branches are suitable for which position. To give respect to branches and flowers, try to use both hands to arrange them. If the stem is too thin to place on *kenzan*, it can be reinforced by binding together with a thicker stem from other plant material (Takenaka 1995).

FIGURE 10.10 *Rikka* style

Source: Photograph by the author.

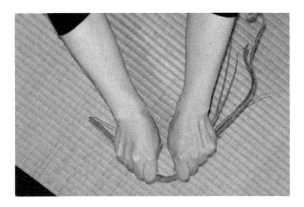

FIGURE 10.11 Squeezing

Source: Photograph by the author.

FIGURE 10.12 *Saika*

Source: Photographed by the author.

Saika

This is a relatively simple arrangement that beginners tend to practice (Figure 10.12). It was designed to express the beauty of a classical *ikebana* style and a western flower arrangement, the shape from *ikebana*, and the color from the European arrangement. The branches and flowers are placed on the *kenzan* collectively, they follow one beautiful, vertical line upwards (see Chart 10.1). This clear one-line style is derived from the classical style.

For this arrangement, we need to prepare several branches and flowers as follows:

Three principal branches

> *Shui*, primary subject branch: the most important line
> *Fukui*, secondary subject: to assist the primary subject
> *Kyakui*, third object: to aid both the primary and secondary subject
> *Kashin*, flowers: arranged before the branches to provide volume and color
> *Nejime*: to hide the *kenzan*

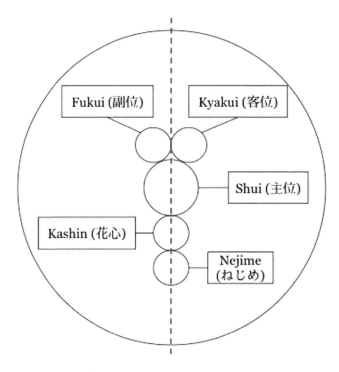

CHART 10.1 *Saika*

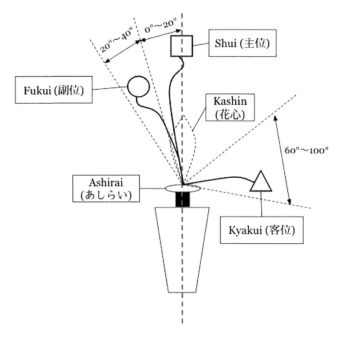

CHART 10.2 *Saika* angle

The three principal branches, *shui*, *fukui*, *kyakui* represent heaven, earth, and mankind.

Length and angle

> *Shui*: the containers diameter plus depth times 1.5
> *Fukui*: 70 percent length of *shui*
> *Kyakui*: 50 percent length of *shui*
> *Kashin*: 30–50 percent length of *shui*
> *Nejime*: 25–30 percent length of *shui*

The *shui* is placed between 0 and 20 degrees from the 90 degree vertical, while the *fukui* and *kyakui* are placed 20–40 degrees and 60–80 degrees from the *shui* respectively (see Chart 10.2). It is encouraged to keep the triangle shape between these branches to maintain the asymmetrical look. Start arranging from the *Shui*, followed by the *Fukui*, *Kyakui*, and others. The branches are hard to place on the *kenzan*, so make sure to use both hands. For a *Saika* arrangement, it is recommended to practice a supplementing technique; if one branch or flower does not have leaves, it can be supplemented by branches with many leaves.

Moribana

This arrangement was created for appreciating the varieties of flowers and colors (Figure 10.13). It was designed during the Meiji period to cherish new western flowers. To emphasize the flowers, the style is emphasized in width rather than in height. For *Moribana*, the three principal branches of techniques, *shui*, *fukui*, and *kyakui* are followed. This style does not follow the one-line policy, thus it is easy to provide width. For expressing the beauty of the *moribana* style, practitioners can practice how to match colors with materials, choosing similar colors is appropriate and ensures the beauty of the *ikebana* arrangement. However, selecting different colors can also provide refined beauty. This divergent color selection might be challenging, but is worth practicing to pursue *ikebana* aesthetics (Kawase 2000.

Three principal branches

> *Shui*: main branch: the most important line
> *Fukui*: secondary branch: to assist the primary subject
> *Kyakui*: third branches: to aid both the primary and secondary subject
> *Ashirai*: flowers or other materials to provide mass, depth, and color

FIGURE 10.13 *Moribana*

Source: Photograph by the author.

Length and angle

> *Shui*: the container's diameter plus depth times 1.5–2
> *Fukui*: 70 percent length of *shui*
> *Kyakui*: 50 percent length of *shui*
> *Ashirai*: 30–50 percent length of *shui*

The *shui* is placed between 20–30 degrees from the vertical, while the *fukui* and *kyakui* are placed around 45 degrees and 60–80 degrees from the *shui* respectively. It is encouraged to keep the triangle shape between these branches to maintain the appearance of asymmetry.

Nageire

This style is a tea flower arrangement influenced by Sen no Rikyū. Since this style emphasizes the natural look, a *kenzan* is not used for stabilizing the arrangement (Figure 10.15). Instead, practitioners are recommended to create support using

FIGURE 10.14 One line in the *kaki*

Source: Photograph by the author.

FIGURE 10.15 *Nageire*

Source: Photograph by the author.

branches inside of the containers. One line, a cross or v–shaped line, is made from branches as supporters at the rim of the container (Figure 10.14). *Nageire* does not follow the rule of the three main branches unlike *Saika*. However, it is desirable to arrange from taller branches to shorter ones. The flower arrangement should be natural, so the branches are arranged behind the flowers. For *nageire* flowers, similar materials are used. This tendency aligns with *chadō* philosophy: emphasizing harmony. Additionally, seasonal flowers are often used for Nageire style; practitioners may select camellia as a winter flower and the Japanese apricot flower as a spring flower to express the transition of seasons in the early spring (Grosser 1997).

Jiyuka *(freestyle)*

Freestyle was established to express the practitioners' creativity (Figure 10.16). Only after learning some basic knowledge is freestyle encouraged. *Ikebana* teacher Sugidate-*sensei* often comments, 'Freestyle is the most difficult. It is

FIGURE 10.16 *Jiyūka*

Source: Photograph by the author.

much easier to follow the rules than express your ideas and image through free-style. Your sense, taste, and aesthetics are expressed through freestyle. This is very challenging.' Grosser (1997) comments that numerous non-Japanese *ike-bana* practitioners are attracted to this freestyle creativity. Some practitioners arrange freestyle *ikebana* for *ikebana* exhibitions; they generally plan to prepare well ahead to examine the theme, design, materials, and utensils. After deciding the rough design, materials, and flower container, the actual *ikebana* is arranged. In order to express the uniqueness, different types of flower materials can be used for freestyle: arranging tropical flowers and classical Japanese branches.

Shinseika

This is a simple arrangement. Practitioners need to prepare only five to seven (ideally five to make it simple) branches and flowers. It follows one line, as in *Seika* style, and *shin, fuku, and negime*. The length of these branches is similar to the *Saika* style above. There are some rules that if bamboo is used as one of the materials, it should be arranged at the front. Flower material examples: 1, *negibozu, ariamu*, sunflower, 2, *okura resuka*, sander sonia.

Shizenkei Shizenka

This is the arrangement to represent nature, so Japanese flowers and plants tend to be used (Figure 10.17). This style uses two *kenzan* on the flat flower

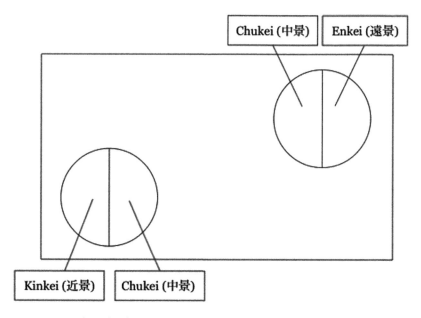

CHART 10.3 *Shizenka* chart

FIGURE 10.17 *Shizenka*

Source: Photograph by the author.

container, suiban style to express far-distance scenery (*enkei*), middle-distance scenery (*chukei*), and close-distance scenery (*kinkei*). For *enkei*, we arrange *shui* and *fukui*, while for the *kinkei*, we arrange the *kyakui*. The length of the *shui*, *fukui*, and *kyakui* are the same as above. The *kenzan* should be hidden by moss.

Ikebana course in higher education

Ikebana courses are offered in higher education in Japan. In order to understand and examine this culture from multiple perspectives, it is desirable to have these courses with both academic discourse and physical *ikebana* practice. To explore flower culture more deeply, the following field trips can be considered.

Field trip

Students are asked to provide a research question related to the field trip and debate it.

Ikebana *exhibition*

Ikebana exhibitions are often held in local towns in Japan. They generally exhibit numerous styles of *ikebana* from classical to creative freestyle. It is indeed worth visiting to be introduced to how different styles of *ikebana* are presented, feel the atmosphere at each exhibition with flower fragrance, and meet *ikebana* practitioners and sense their passion. Students can also begin to get a feel of *ikebana* practitioners' backgrounds in relation to gender and class issues. There are some *ikebana* study groups overseas, and if this course is held outside of Japan, it might also be worth contacting them. It is also worth visiting western flower arrangement shows to feel the differences from *ikebana* aesthetics.

Museum

To explore the aesthetics such as *wabi sabi* and *miyabi* concepts, it is also beneficial to visit local museums. Students can experience these aesthetics in other art forms: paintings, pottery, and calligraphy. Some museums outside of Japan, such as the British Museums, Asian Art Museum in San Francisco, and Seattle Asian Art Museum also present exquisite artworks related to these aesthetics.

Visiting other traditional art forms

Japanese Dance, *noh*, *chadō* gatherings: Exploring other art form gatherings is helpful to examine the similarities or differences with *ikebana* in relation to the *iemoto* system, and the relationship among practitioners and the rituals. The *ikebana* course in Akita International University, AIU, went to watch a Japanese Dance performance. They were able to watch and recognize specific movements in Japanese Dance and also learn of the hierarchy systems among practitioners.

Japanese gardens

Visiting Japanese gardens allows students to feel the garden beauty using their five senses: the sound of wind, water, and gravel, the scenes of greenery, and the smell of the moss and leaves. AIU *ikebana* courses visited a local Japanese garden, Jyoshitei owned by Lord Sakate, and were introduced to how flowers had been arranged in the garden. As described in the garden chapter, there are numerous Japanese gardens in the world, so it is worth searching for and visiting one or more of these gardens.

Local artist/artisan's work

Ikebana uses a variety of containers. It is therefore interesting to visit local kilns or art studios to appreciate the efforts that artists put into creating a single piece of art. Courses can visit local studios specializing in lacquer ware, pottery, and bamboo crafts as well as having the opportunity to conduct interviews with the artisans themselves. It might also be worth trying to make one of the art pieces, as many of the studios offer trial opportunities.

Online courses

There has been a significant demand to teach *ikebana* courses online due to the COVID-19 pandemic. Teaching *ikebana* practice online is also possible. Students are requested to prepare several utensils: the flower containers (*kaki*) *and* some scissors and flowers from their local town. A teacher can first guide how to arrange a specific *ikebana*, and then students can practice and arrange flowers by themselves. Guidance can be offered through a webcam. Make sure that the teacher can see students' arrangement through appropriate angles by offering enough distance from students' webcam and having a clear background behind their *ikebana* arrangements. After completing their arrangements, students can take their pictures and upload them to an online shared folder where the teacher can instantly view and comment on their work during a class.

If the *ikebana* courses have been presented during the time when students found it difficult to acquire flowers and branches due to their quarantine time, and city lock down situations, etcetera, the course can offer to practice *ikebana* with whatever they can acquire: flowers from their own garden, dry flowers from their home, or even a cardboard box painted green with tissue paper flowers (Figure 10.18). The most important philosophy is to move their hands, and not just observe through their screen. Japanese traditional arts often instruct the practitioner not to remember through their minds, but through their bodies (*karada de oboeru*). I strongly encourage students to practice *ikebana* with whatever they have!

Research question

If you are aiming for *wabi sabi*, is it wrong to use wire?

FIGURE 10.18 Student arrangement with carbon box

Source: Photograph by the author.

Conclusion

This chapter shows how *ikebana* practitioners focus not only on flowers, but on containers and other related techniques. It also explores how *ikebana* can be creative, particularly for freestyle, while many Japanese traditional arts are considered to be not creative. Freestyle does not have any restrictions; even the seasonal arrangement is optional. This may be the reason why *ikebana* is well accepted in other countries. This creativity in *ikebana* appears to have great potential for the future of *ikebana*, as it provides a welcoming atmosphere for practitioners to express their own individual ideas and aesthetics. This important aspect of *ikebana* might become more widely recognized as Japanese education and society continue to encourage and help shape an individual's creativity.

References and further reading

Davey, H. E. (2000) *The Japanese Way of the Flower: Ikebana As Moving Meditation Michi, Japanese Arts and Ways*, Berkeley CA: Stone Bridge Press.

Fujiwara, Y. (1976) *Rikka, The Soul of Japanese Flower Arrangement*, Tokyo, Japan: Shufunotomo, LTD.

Grosser, R. (1997) *10 Lessons in Ikebana*, Queensland: Australian Ikebana Centre.

Ikenobo Kadokai. (2013) *Hana no Arakaruto*. No. 54, Kyoto: Ikenobo Kadokai.

Kawase, T. (2000) *The Book of Ikebana*, Tokyo, Japan: Kodansha.

Lim, L. (2009) *Essential Ikebana*, Singapore: Marshall Cavendish International.

Norman, S. (1991) *Ikebana: 101 Plants and 478 Arrangements*, New York: Weatherhill.

Okada, K. (1992) *Ikebana with the Seasons*, Tokyo, Japan: Shufunotomo, LTD.

Sparnon, N. (1991) *Ikebana 101 Plants and 478 Arrangements*, Tokyo, Japan: Shufunotomo, LTD.

Takenaka, R. (1995) *Japanese Flower Arrangement IKEBANA Step By Step*, Tokyo: JOIE.

——— (1996) *Enchanting Ikebana: Japanese Flower Arrangement*, Tokyo: Japan Publications Trading Co.

Teshigahara, H. (1997) *The Art of Ikebana*, New York: Weatherhill.

Teshigahara, W. (1990) *Ikebana: A New Illustrated Guide to Mystery*, Tokyo, Japan: Kodansha.

11
FUTURE

As discussed in Chapter 3, History, there is an apparent fact that the number of practitioners of *ikebana* is decreasing drastically due to the lack of interest from the younger generations. Approximately 4.5 million people were practicing *ikebana* in 1996, which then decreased to about 2 million in 2016 (Hamasaki 2021). Most practitioners in Japan are older women and 78 percent of *ikebana* practitioners are over 60 years old (Hamasaki 2021). This chapter explores the changes made to traditional customs and other attempts to promote *ikebana* and other traditional arts to attract younger generations in Japan. It will also introduce overseas activities which promote *ikebana* worldwide. This chapter further draws a summative conclusion of this book as a whole.

Ikebana international activity

As outlined in Chapter 8, 'Gender and class' Ikenobo introduced *ikebana* to foreign women during the Allied Occupation of Japan after World War II. Stalker (2018) states that 95 percent of non-Japanese national women learned *ikebana* and they introduced it to their home countries upon their return. Due to these foreign women, *Ikebana* International was founded in 1956 to promote *ikebana* abroad. There are currently 143 branches across the world and over 7,000 members (*Ikebana* International 2021). Additionally, the Northern California Chapter, NCC, based in San Francisco, was officially founded in 1963 and has practiced *ikebana* ever since. Moreover, the NCC has offered numerous *ikebana* demonstrations to local communities. Not only does the NCC offer regular practice sessions and demonstrations, practitioners also offer *ikebana* lessons to local schools to encourage fostering future *ikebana* practitioners and teachers. Aside from the activities at *Ikebana* International and the NCC, Ikenobo has opened

DOI: 10.4324/9781003248682-12

around 100 branches, Ohara, 64 branches, and Sogetsu around 120 branches to practice (The Northern California Chapter 2021).

IKENOBOYS

To attract different backgrounds of people, including youths, adolescents, and males, Ikenobo has promoted male *ikebana* groups known as IKENOBOYS since 2016. Ikenobo named this group IKENOBOYS implying they were '*hana wo ikeru men,*' (men who arrange flowers) and '*ikemen*' (handsome men) (Ikenobo Kadoka 2021). They promote *ikebana* in shopping centers, and via media platforms. IKENOBOYS also promote individual flower arrangements, as well as workshops to companies and organizations. For instance, Ikenobo provided the *ikebana* arrangement at the *sumo* and *kimono* ceremonies. They also provide workshops for hotel companies to learn the concepts of hospitality through flower arrangements.

As can be seen in Figure 11.1 below, this book aims to shift the norm from *ikebana* being for middle class women or girls to a more inclusive and less gendered art form. The first IKENOBOYS group was organized with *ikebana* teachers and practitioners, but the new IKENOBOYS were organized in 2020 with 12 male practitioners with a range of occupations and from a variety of places in Japan (Ikenobo Kadoka 2021). IKENOBOYS groups attract teenagers and young Instagram users, and the official Instagram account currently has over 5,000 followers (Ikenobo Kadoka 2021).

FIGURE 11.1 IKENOBOYS. Copyright Ikenobo Kadoka

Ikebana and digital art

Recently, *ikebana* exhibitions have done numerous collaborations with digital technology. *Ikebana* Ryusei school conducted an *ikebana* exhibition for its 130th anniversary with a digital rock garden. Various types of digital rock garden designs were displayed in the center of the exhibition (Ryusei 2021). This school has also organized interactive *ikebana* exhibitions with CG shadow. A CG bird's shadow, flying around the *ikebana* arrangements' shadow, stopped and chirped on branches. When the audience's shadow touched the *ikebana*'s shadow, it synthesized with the shining light (Iwasaki *et al.* 2017). On the other hand, Ikenobo also provided an *ikebana* arrangement with digital art in Kyoto in 2019 and 2020. The next Ikenobo *Iemoto* arranged *ikebana* through digital art with the image of a phoenix, which represents eternal prosperity. In 2020, at Nijyo castle, flowers were arranged with illuminations, which expressed the wind. Two members of IKENOBOYS, Kosugi and Baba, arranged this *ikebana* to express the beauty of petals, leaves, and bamboo all flowing with the wind (Ikenobo Kadoka 2021) (Figure 11.2).

Digital technology is also introduced for daily practice. Cado, the application for tablet computers, makes designs based on flowers and guides arranging materials by following *ikebana* aesthetics (Yokokubo and Shio 2014). The TracKenzan computer application has also been introduced as *ikebana* training

FIGURE 11.2 *Ikebana* and illumination in Kyoto. Copyright Ikenobo Kadoka

material. TracKenzan, the trackpad, and the sylus are used as a *kenzan* and a flower axis respectively to create 3D CG *ikebana* arrangements (Yokokubo *et al.* 2019). Yokokubo *et al.* (2014, 2019) state that these applications open more opportunities to engage with *ikebana*, especially for younger generations.

Sound of ikebana

The technology side of Japan also pays attention to *ikebana* beauty. This beauty was expressed through cutting-edge technology by Naoko Tosa and presented at Times Square in New York (Figure 11.3). In conjunction with Times Square Art, and employing over 60 buildings, Tosa's display expressed the beauty of *ikebana* through sound and splashes of color flying in slow motion across dark screens using over 60 buildings (Mai *et al.* 2020). This slow-motion liquid was filmed at 2,000 frames per second and it expresses the asymmetrical *ikebana* aesthetics imagining spring. Tosa argues (2021) that *ikebana* can be developed as a painting tool in digital art by applying AI techniques. Tosa further made a project to combine the *Sounds of Ikebana* to work with the first sounds of a newborn baby to celebrate the baby's birth. Working with *ikebana* Misho school *iemoto*, Tosa also expressed the decayed beauty (*kuchi no bi*) by capturing the moment when the frozen *ikebana* arrangement was destroyed by an air gun. Misho school *iemoto* states that this decayed beauty is also considered a precious aesthetic. She further states that culture, including traditional arts, is still alive in modern society; her interpretation of *ikebana* beauty often seems to attract new and different audiences.

Ikebana and media

Traditional art has been perceived as high culture, but it has also started to engage with the pop culture field, TV drama. A Nippon Television Network TV drama about *ikebana iemoto*'s daughter premiered in 2018. It was a story about a daughter of the *iemoto* of an *ikebana* school who fell in love with a bicycle repair shop man. One of Japan's most popular actresses, Satomi Ishikawa, performed as the daughter of the *iemoto*. The show discusses *iemoto*'s duty, love, and relationship

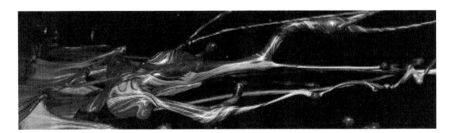

FIGURE 11.3 *Sound of Ikebana.* Copyright Tosa Naoko

issues from different social backgrounds, and also the beauty of *ikebana* (Nippon Television Network Corporation 2021). Critics who reviewed the show said that it was not as popular as they had expected it would be, however, it appeared that it did provide a guide for the younger generations to feel that *ikebana* culture might be closer to them than they realized. Similar to this *ikebana* related TV drama, *noh iemoto*'s story premiered by TBS (Tokyo Broadcasting System Television) in 2020. The director Kudo wanted to state the issue of successors in the traditional arts field. This drama comically describes the *iemoto*'s son's struggle for pursuing his passion to be a professional wrestler or becoming the *noh iemoto* and carrying on his family duty and legacy (Tokyo Broadcasting System Television, Inc 2021). On the other hand, Sakabe (2014) stated that another TBS TV drama about *kabuki*, *Tiger and Dragon*, had triggered the young generation viewers to visit and see *kabuki* performances at the theater. Several of my students commented that these TV dramas made them want to take traditional art courses.

Ikebana and education

As described in the previous chapter, *ikebana* tends to be taught less widely than other subjects. On the other hand, Sato (2020) comments that teachers in the compulsory education system struggle to find the class topics in English and moral education classes. Sato (2020) comments that this traditional art education in English and morals education classes can be one possible way to teach traditional art in the compulsory educational system. According to Yoshimura (2013), at a school in Nara prefecture, *noh* is focused on in English class in elementary school; students learn the *noh* movement with basic English. Several elementary schools including a school in Akita prefecture have offered *ikebana* lessons as one of the ways to learn plants in science classes. By collaborating with a local *ikebana* teacher, this unique way of introducing *ikebana* is possible.

While most higher education institutions offer *ikebana* as part of their extra-curricular activities, some academic institutions have offered *ikebana* courses as academic credit-bearing courses. There is an increase in demand for academic credit-bearing *ikebana* courses due to the increasing number of international students, combined with Japanese students who have more opportunities to learn and integrate knowledge of their own culture within a global context. Most of the institutions offer this course for Japanese or international students. This course can be offered for both Japanese and international students as an intercultural collaborative learning style enhances their creative and active learning through *ikebana* (Suematsu 2020).

Ikebana as healing

Flowers and plants are sometimes recognized as a therapeutic option in our stressful society. Kaku (2017) and Kimura *et al.* (2012) argue that viewing Japanese gardens relieves viewers' stress and anxiety. Japanese garden viewing also appears to improve viewers' physiological issues including muscle stiffness (Kimura *et al.* 2012) and heart rate issues (Goto *et al.* 2016). Similar results have been reported for *ikebana* arrangements. Lee *et al.* (2012) states that flower arranging can be used for physical rehabilitation for stroke patients. Mochizuki (2016), shares that *ikebana* was effective in helping some people heal from mental stress which arose after the great earthquake in 2011 in Japan. Mochizuki (2016) explains that the number of people who had physical symptoms of stress such as headaches and not being able to sleep decreased after *ikebana* experiments. Homma *et al.* (2015) state that *ikebana* practice can slow one's respiratory rate and has been used as a remedy for anxiety. *ikebana* is also utilized for aiding dementia patients (Hamazaki 2000, Ikenobo *et al.* 2015). Ikenobo *et al.* (2015) offered *ikebana* lessons to elderly people who suffer from Poriomania and Alzheimer's. The experimental study indicated that *ikebana* practice can reduce emotional instability, loss of motivation, insecurity, and anxiety more than watching DVDs and doing karaoke activities. Ikenobo *et al.* (2015) further stated that appreciating a sense of seasonal materials, using the five senses by touching or seeing, using hands with tools, and interacting with other people are possible contributing factors for this therapy.

Invention of tradition; new *iemoto* system

So far, we had a look at *ikebana*'s activities to attract more diverse audiences. It appears that *ikebana* has actively begun to adopt the new social trends. What is apparent is that even the most renowned Ikenobo schools are accepting the new trends and adjusting themselves according to the shifts in society. It seems to be significantly different from other traditional art fields in Japan. For instance, the Ikenobo website does not provide license costs, yet it provides relative information in terms of cost and time of flower arrangements provided by *ikebana* professionals. It also offers SNS access sites to provide up-to-date information. Further, it offers contact details to the public without emphasizing a certain level of formality by using the phrase 'Okigaruni (without hesitating)' (Ikenobo Kadoka 2021). The *iemoto* system is also questioned and amended in the *ikebana* world. Kasumi, the designated heir of Teshigahara Sofu of the Sogetsu School, vowed to change the system to inhibit excessive profits, and criticized the hierarchical structure of the system that provided the most benefit to the patriarch. The daughter of Adachi school, Toko also criticized the hereditary *iemoto* and issued a public 'declaration of independence' from her father by earning a reputation based on her own talents (Stalker 2018).

More than anything else, the *ikebana iemoto* system now offers the position of *iemoto* (grand master), to women and provides a sense of equal opportunities and gender equality. Other traditional arts such as *noh* are also offering more opportunities to women by allowing females to make *noh* masks. However, the availability of new positions and responsibilities for women are much more limited compared to *ikebana*.

Conclusion

This book examined *ikebana* and flower culture from anthropological and socio-logical perspectives. It analyzes discussions including Japanese aesthetics, customs and rituals related to flower arranging, *ikebana* history and the *iemoto* system. *Ikebana* and its surrounding system seem to never be the same throughout history; it has always evolved. *Ikebana* for middle class women's image is still strong since the Meiji period in Japan, yet, this image might also change in 30 years. On the other hand, flowers still appear to be appreciated as a tool to convey our real feelings in Japan today and in the future in relation to the Japanese concept of real feelings and public faces (*honne* and *tatemae*). Kawabata Yasunari (1899–1972) gave a speech, *Japan, the Beautiful and Myself*, at the Nobel Prize awards ceremony in 1968. He commented about the in-depth beauty in Zen, *chadō*, and *ikebana* and stated that 'The single flower contains more brightness than a hundred flowers.' It appears that this kind of appreciation of simple flowers can still be seen.

With the COVID-19 pandemic, staying home has been encouraged in Japan for over two years. People have started to have more opportunities to engage with flowers and gardening at home. Fewer flower arrangements are sold in the business field, flower companies are targeting the individual to receive delivered flowers straight to their homes: monthly delivery of flower arrangements at home from 500 yen. These changes might provide more closeness to flowers in our daily lives. How about *ikebana*? Will it survive? Considering the adjustments, it appears that, more than other traditional arts, *ikebana* might have the potential to survive, particularly considering promotions to different genders and generations. From analyzing gender inequalities in the cultural sector, Pujar (2016) states that to change gender roles, which have been deeply ingrained in societies for centuries, is considerably difficult. At least, *ikebana* in the Japanese traditional arts world appears receptive to welcoming impartial and open perspectives on gender roles. Yet, how about the new generations?

In this rapid society, this traditional art, *ikebana*, which emphasizes slowness, tends to be forgotten. Cross (2009) states that these traditional arts are no longer popular for the younger generations who seek immediate and efficient know-ledge. 'Stop and smell the roses.' The Japanese do not have this saying, but they might need this saying in Japanese society today and going into the future. On the other hand, we do see adjustments to this slow traditional art image, par-ticularly concerning the future generations involved in *ikebana*. Not only has it become important to make the qualification system faster and clearer than before, but *ikebana* culture is also incorporating different flower arrangement styles to suit new audiences. However, another argument will be: Do we need to change *ikebana* culture into a fast culture? Do we have to adjust to efficient learning and offer a quick qualification? We could argue that just maintaining current *ikebana* culture at its current slow pace is just as important.

Having said that, there is an apparent fact that *ikebana* practitioners are decreasing drastically. How many practitioners will we have in 30 years? It appears that *ikebana* might need to change its image from *ikebana* for middle-class women to *ikebana* for every gender, every nationality. However, would this be enough elements to let *ikebana* culture survive? Certain government financial support in relation to education, medical or healthcare related projects with *ikebana* seems to be necessary to keep this culture alive. As mentioned above, *ikebana* has a great potential to contribute to maintaining the healthy mental condition of both senior residents, who are the majority in Japan, and those who suffer from mental illness, perhaps from the rapid and efficient mode of society that most of us are trying to promote. The flowers and plants that we use for *ikebana* are indeed delicate, yet some of them are tough. I believe *ikebana* has the vitality to endure, to flourish as dandelions delighting in spring's sweet incandescence.

References and further reading

Aida, A., Chang, K., Suzuki, M., and Taniguchi, S. (2003) 'Research on image of healing received from garden landscape,' *Journal of Agriculture Science, Tokyo University of Agriculture*, 48(3), 115–127.

Carrasco, R., Lacertosa, T., and McCastle, A. (2017) *Physiological Responses to Activity by Novice vs. Advanced Ikebana Practitioners*, Nova Southeastern University Works. Retrieved from https://nsuworks.nova.edu/ot_colloquium/third/events/9/ (accessed 3 February 2021).

Cinco, K., Carrasco, R., and Key, C. (2016) *Relationship between Anxiety, Stress-Response, and Lived Experiences Post Ikebana Intervention: A Pilot Study*, Nova Southeastern University Works. Retrieved from https://nsuworks.nova.edu/ot_colloquium/second/events/5/ (accessed 3 February 2021).

Cross, T. (2009) *The Ideologies of Japanese Tea: Subjectivity, Transience and National Identity*, Global Oriental: Folkstone.

Goto, S., Gianfagia, T. J., Munafo, J. P., Fujii, E., Shen, X., Sun, M., Shi, B. E., Liu, C., Hamano, H., & Herrup, K. (2016) 'The power of traditional design techniques: The effects of viewing a Japanese garden on individuals with cognitive impairment' *HERD: Health Environments Research & Design Journal*, 10(4), 74–86. DOI: 10.1177/1937586716680064

Hamazaki, E. (2020) 'Ikebana kaido kaisai ni okeru ninchisho no hito no ikebana ryoho niyorushakai sanka no koka' [The effects of social participation by 'ikebana therapy' in the event of 'ikebana Street' of arranged flower with dementia people], *Doshisha University Policy & Management Review*, 22(1), 63–77.

——— (2021) 'Kado no genjyo to kadai ni taisu ikebana ryoho gainen no kakushinsei ni kansuru jitsuen teki kousatsu' [A study of ikebana therapy through investigating contemporary issues in kado], *Institute for the Study of Study of Humanities and Social Sciences, Doshisha University. Shakai Kagaku*, 50(4), 147–175.

Homma, I., Oizumi, R., and Masaoka, Y. (2015) 'Effects of practicing ikebana on anxiety and respiration.' *Journal of Depression & Anxiety*, 4, 1–5.

Ikebana International (2021) *Ikebana International Vol 63*, Tokyo: Ikebana International.

Ikenobo Kadoka (2021) Personal communication 17 June.

Ikenobo Y., Kida Y., Kuwahara N., Goto A., and Kimura A. (2013) 'A study of the effect of the shape, the color, and the texture of ikebana on a brain activity' in: Duffy V. G. (eds.), *Digital Human Modeling and Applications in Health, Safety, Ergonomics, and*

Risk Management. Human Body Modeling and Ergonomics, Berlin: Springer, pp. 59–65. DOI: /10.1007/978-3-642-39182-8_7

Ikenobo, Y., Mochizuki, Y. and Kuwahara, A. (2015) 'Usefulness of ikebana a nursing care environment,' in Duffy V.G (eds.) *Digital Human Modeling. Applications in Health Safety, Ergonomics and Risk Management: Ergonomics and Health.* Berlin: Springer, pp. 441–447.

Iwasaki, H., Mizuno, S., and Akiba Y. (2017) 'Interactive digital contents with ikebana and CG, "Digital Karesansui" and "Shadow Picture of Ikebana",' *Information Processing Society of Japan*, 64(1), 29–31.

Kaku, R. (2017). 'Arutsuhaima kanja nihon teien wo mite otitsuku' [Alzheimer's patient calms down by looking at Japanese gardens], *Reuters*. Retrieved from https://jp.reut ers.com/article/idJP00093300_20170809_01920170809 (accessed 3 February 2021).

Kawabata, Y. (1968) *Japan, the beautiful and myself. The Nobel Prize.* www.nobelprize.org/ prizes/literature/1968/kawabata/lecture/ (accessed 14 October 2021).

Kimura, T., Matsumoto, K., Okada, Y., Uchida, S., and Yamaoka, J. (2012) 'Psycho-physiological effects by the appreciation of the garden and the art,' *Research Reports from the MOA Health Science Foundation*, 31–39. DOI: 40019547360

Kwon, H. (2016) 'Gendai kankou ni okeru dentoubunka to popyura- karuchya-tottori-ken sakaiminato-shi to nagano-ken ueda-shi no kankou ni mirareru bunka no datsubunmyakuka' [Traditional Culture and Popular Culture in Contemporary Tourism: Cultural Decontextualization in the Tourism of Sakaiminato-shi and Ueda-shi], *Journal of Tourism, Kankogakuron* 42(2), 121–133.

Lee, S., Park, S., Known, O., Song, J., and Son, K. (2012) 'Measuring range of motion and muscle activation of flower arrangement tasks and application for improving upper limb function,' *Horticultural Science & Technology*, 30(4), 449–462.

Mai, C. H., Nakatsu, R., and Tosa, N. (2020) 'Developing Japanese ikebana as a digital painting tool via AI,' in Nunes, N. J., Ma, L., Wang, M., Correia, N., Pan, Z. (eds.), *Entertainment Computing. ICEC 2020.* Berlin: Springer, pp. 297–307.

Mizuno, S. and Yoshimura, S. (2020) 'Dejital eizo wo mochiita atarashii ikebana hyogen no sozo ni kannsuru kenkyu' [Research on the creation of new ikebana expressions using digital images], *Bulletin of Research Institute for Industrial Technology, Aichi Institute of Technology*, 22, 1–7.

Mochizuki, H. (2016) 'Structured floral arrangement program developed for cognitive rehabilitation and mental health care,' *Japan Agricultural Research Quarterly*, 50(1), 39–44.

Mukai, N., Takara, S., Segawa, A., Nakagawa, M., and Kosugi, M. (2010) 'Prototype system for ikebana exercise with haptic device.' *The Institute of Image Information and Television Engineers*, 64(10), 1510–1515. DOI:10.3169/itej.64.1510

Nippon Television Network Corporation (2021) *Takane no Hana* (Flower that cannot reach), www.ntv.co.jp/takanenohana/ (accessed 14 October 2021).

Pujar, S. (2016) *Gender Inequalities in the Cultural Sector*, Culture Action Europe.

Ryusei (2021) *Ikebana Ryusei September 2021 issue*, Tokyo: Ryusei Kadokai.

Sakabe, Y. (2014) '"Dentougeinou" no ima: Sengokabuki, rakugokougyou no keiryoubunseki kara.' [The latest traditional performing arts: From the econometric analysis of postwar kabuki and rakugo performances]. Ritsumeikan University Thesis. DOI: 10.34382/0001015

Sasaki, M., Oizumi, R., Homma, A., Masaoka, Y., and Ijumi, M. (2011) 'Effects of viewing ikebana on breathing in humans,' *Showa University School of Medicine*, 23, 59–65. DOI:10.15369/sujms.23.59

Suematsu, K. (2020) 'Intercultural collaborative learning in extracurricular activities: characteristics and students' learning and development,' *Journal of International Student Education* (25), 9–20.

Stalker, N. (2018) 'Flower empowerment: Rethinking Japan's traditional arts as women's labor,' in Bullock J., Kano A., & Welker J. (eds.), *Rethinking Japanese Feminisms*. Honolulu: University of Hawai'i Press, pp. 103–118.

The Northern California Chapter. (2021) *San Francisco Bay Area Chapter* www.ikebana.org/ikebana-for-youth/ (accessed 14 October 2021).

Tokyo Broadcasting System Television, Inc (2021) *Ore no Ie no hanashi* (The Story of My Home), www.tbs.co.jp/oreie_tbs/ (accessed 14 October 2021).

Tosa, N. (2021). Personal Communication November 29.

Yokokoubo, A. and Siio, I. (2014) 'Cado mijikana kazai wo riyoshita iebana shien system' [CADo: Supporting System for Flower Arrangement], *Information Processing Society of Japan*, 55(4), 1246–1255.

Yokokubo, A. (2019) 'TracKenzan: Trakkupaddo to tacchioen wo mochiita ikebana rensyu sistemu' [TracKenzan: An ikebana practice system using a trackpad and stylus], *Information Processing Society of Japan*, 59, 2006–2018.

GLOSSARY

Ikebana flower

airisu: iris
ajisai: hydrangea
akebi: akebia chocolate vine
anemone: anemone
ansuryumu: anthurium
asagao: morning glory
asuparagasu: asparagus
ashi: reed
awa: foxtail
ayame: iris sanguinea
azami: plumed thistle

baimo: fritillary
bara: rose
bashō: Japanese banana / basjoo
begonia: begonia
benibana: safflower
biwa: roquat
boke: Japanese quince
botan: peony

chūrippu: tulip

daria: dahlia
dendorobyumu: dendrobium

endō: pea

fuji: wisteria
fujibakama: boneset
fukinotō: Japanese butter-bur
fukujyuso: amur adonis
furijia: freesia
futoi: Japanese bulrush
fuyō: cotton rose

gābera: gerbera
gama: great cat's tail
gibōshi: plantain lily
gurajiorasu: sword lily

ha-botan: decorative kate
hagi: bush clover
haibisukasu: hibiscus
hamanasu: sweet brier
hana-mizuki: flowering dogwood
hana-shōbu: Japanese iris
haran: barroom plant
hasu: lotus
higanbana: red spider lily (*lycoris radiate*)
hiiragi: holly olive / false holly
himawari: sunflower
hinagiku: daisy
hirugao: wild morning glory
hitori-shizuka: cloranthus japonicus
hiyashinsu: hyacinth
hotaru-bukuro: bell-flower (*campanula punctata*)
hototogisu: toad-lily (*tricyrtis hirta*)
hōzuki: Chinese lantern plant
hyakunichisō: common zinnia

jasumin: jasmine

kaede: maple
kakitsubata: rabbit-ear iris
kānēshon: carnation
karā: calla / calla lily
karasuuri: snake gourd
kashiwa: daimyo oak

kasumisō: baby's breath / gypsophila
katakuri: dog's-tooth lily
katoreo: cattleya
kawara-nadeshiko: superb pink (*dianthus superbus*)
keitō: cockscomb
keshi: poppy
kiku: chrysanthemum
kikyō: balloon flower
kingyosō: snapdragon
kinsenka: pot marigold
kinshibai: hypericum patulum
kiri: paulownia
kirinso: goldenrod
kobushi: northern Japanese magnolia
kochōran: butterfly orchid
kodemari: reeves spirea
kosumosu: cosmos
kuri: chestnut
korokkasu: crocus
kuchinashi: cape jasmine
kuremachisu: clematis
kuromoji: spicebush
kuzu: kudzu vine

manryō: coral ardisia (*ardisia crenata*)
matsu: pine
miyakowasure: "miyakowasure" (aster savatieri)
mokuren: lily mongolia
momiji: Japanese maple
momo: peach blossom
mokuge: rose of Sharon
murasakishikibu: Japanese beauty-berry

nadeshiko: superb pink
nanakamado: mountain ash
nanohana: field mustard / flower brassica
nanten: nandina / heavenly bamboo
neko-yanagi: Japanese pussy willow (*salix gracilistyla*)
nichinichisō: madagascar periwinkle

ōbai: winter jasmine
okatorano'o: gooseneck
ōyama-renge: Siebold's magnolia

panjī: pansy
poinsettia: poinsettia
popī: poppy

rabendā: lavender
rairakku: lilac
ran: orchid
rindō: gentian

sakura: cherry blossom
shakunage: rhododendron
shakuyaku: Chinese peony
shidare-yanagi: weeping willow
shidare-zakura: weeping cherry
shiran: bletilla
shūmei-giku: Japanese anemone
suiren: water lily
suisen: angel's teas' daffodil / narcissus
suītopī: sweetpea
sumire: violet
susuki: eulalia / pampas grass
sutokku: stock
suzuran: lily of the valley

take: bamboo
tanpopo: dandelion (*taraxacum*)
tessenn: clematis
torano'o: speedwell / gooseneck
toruko-gikyo: prairie gentian
tsubaki: camellia
tsukimisō: evening primrose
tsukushi: field horsetail (*equisetum arvense*)
tsuta: ivy
tsutsuji: azalea
tsuyukusa: day flower (*commelina communis*)

ume: Japanese apricot

warabi: bracken
wasurenagusa: forget-me-not

yama-ajisai: mountain hydrangea
yamabuki: kerria
yamagi: willow

yūgao: moonflower (*lagenaria siceraria*)
yuri: lily

zakuro: pomegranate
zenmai: osmund

Tea flower

amadokoro: Solomon's seal

ebine: calanthe
enokoro-gusa: foxtail grass
ezo rindo: gentian

fuji: wisteria
fuji-bakama: mistflower
fuji-utsugi: butterfly bush
fuki-no-to: flower-of-the-butterbur
fushiguro-senno: lychnis

hagi: bush-clover
hakone utsugi: weigela
haku' unboku: storax, snowbell
hama-endo: beach-pea
hana-ikada: raft flower
hangesho: lizard's tail
hashibami: hazel
higan-zakura: equinox cherry
hime-yuri: morning star lily
hirugao: bindweed
hitori-shizuka: chloranthus
hochaku-so: fairy bells
hokyoji nadeshiko: hokyoji pink
hotaru-bukuro: bell flower

inaka-giku: aster
iwakagami: shortia

jinchoge: daphne

kaido: apple bough
karaito-so: burnet
karamatsu-so: meadow rue
kawara-nadeshiko: pink

kayatsuri-gusa: umbrella sedge
keitou: cocks comb
ki-bushi: tree wisteria
kikyo: balloon flower
kin-mizuhiki: agrimony
kobushi: cucumber tree
kodemari: little handball
kumagai-so: lady's slipper
kuro-yuri: black lily
kuromoji: spicebush
kusa boke: flower grass quince
kuzu: arrowroot, kudzu vine

maizuru-so: Greek mayflower
mansaku: witch-hazel
matsuyoi-gusa: evening primrose
mayumi: spindle tree
miso-hagi: longpurple, willowweed
mizo-soba: knotweed, smartweed
mizu-basho: swamp basho
mizuhiki: smartweed, knotweed
mukuge: shrubby althea, rose of Sharon
murasaki-shikibu: Japanese beautyberry

nanohana: rape flower, the flower of greens
natsu-haze: bilberry
natsu-tsubaki: summer camellia
neko-yanagi: pussy willow
nishikigi: winged spindle tree
no-azami: thistle
nokon-giku: aster
numa torano-o: loosestrife-of-the-swamp

o-tade: prince's father
odamaki: columbine
odoriko-so: dead nettle
oka torano-o: loosestrife-of-the-field
okina-gusa: pasque flower
ominaeshi: patrinia
oni-yuri: tiger lily
oyama rindo: gentian

rindo: gentian

sanshuyu: sanshu
shidare-yanagi: Napoleon's willow narcissus
shima-ashi: striped bog-reed
shimotsuke: spirea
shukaido: Chinese begonia
suiren: water lily
suisen: French daffodil, narcissus
susuki: eulalia, Japanese pampas grass

tessen: clematis
tsubaki: camellia
tsukinuki-rindo: trumpet honeysuckle
tsurifune-so: touch-me-not, jewelweed
tsuru-umemodoki: bittersweet, staff tree
tsuwabuki: farfugium
tsuyu-kusa: day flower

uguisu-kagura: honeysuckle
ume: plum
umebachi-so: grass-of-Parnassus, white buttercup
unohana: mock orange
unryu-yanagi: cloud-dragon willow

wabi-suke: Judith camellia
waremokou: burnet-bloodwort

yama-boshi: dogwood
yama-shakuyaku: Japanese peony
yama-zakura: Japanese flowering cherry
yamabuki: Japanese yellow rose
yamabuki-so: celandine poppy
yomena: aster, field chrysanthemum
yukiwari-so: primrose

zakuro: pomegranate
zazen-so: meditating flower

INDEX